26 MAY 1987
AF878893

HOW MACHINES THINK

A GENERAL INTRODUCTION TO ARTIFICIAL INTELLIGENCE

ILLUSTRATED IN PROLOG

Nigel Ford
University of Sheffield, UK

JOHN WILEY & SONS
Chichester · New York · Brisbane · Toronto · Singapore

POLYTECHNIC SOUTH BANK
CLASS No. 006.3 FOR
DATE 5/87
ACC. NO. 34491160
L/R L LOAN

Copyright © 1987 by John Wiley & Sons Ltd.

All rights reserved.

No part of this book may be reproduced by any means, or transmitted, or translated into a machine language without the written permission of the publisher

Library of Congress Cataloging-in-Publication Data:
Ford, Nigel
How machines think.

Bibliography: p.
Includes index.
1. Artificial intelligence. 2. Machine learning.
I. Title.
Q335.F67 1987 006.3 86–15953
ISBN 0 471 91139 9

British Library Cataloguing in Publication Data:
Ford, Nigel
How machines think : a general introduction to artificial intelligence.
1. Artificial intelligence
I. Title
006.3 Q335

ISBN 0 471 91139 9

Typeset by Photo·graphics
Printed and bound in Great Britain

This book is dedicated to
my mother and father, and to
Rosalind, Dominic, and Alasdair

CONTENTS

Introduction		1
Part A: Basic processes		5
1	Storing knowledge	7
2	New knowledge for old	15
3	Spreading diseases and tracing relations: recursion	21
4	Do it yourself	33
Part B: Strategies for intelligent behaviour		53
5	Putting your (mathematically) best foot forward	55
6	Plans and subplans	63
7	More sophisticated strategies	71
8	Probabilities	80
9	Frames of reference and expectations	90
10	A simple blackboard system	101
11	Learning	118
Part C: Intelligent systems and their development		131
12	Working systems	133
	STRIPS: means–ends analysis	133
	ABSTRIPS: planning at separate levels	136
	MOLGEN: planning at interacting levels	139

SAVOIR: dealing with probabilities 141
HEARSAY: blackboard techniques 144
EXPERT-EASE: learning by induction 147
AM and EURISKO: automated discovery using frames 150
ACT: modelling the human mind 154
13 Knowledge-engineering tools 158

Part D: Computers as intelligent life? 171

POSTSCRIPT 185

Appendix 1 Programming in PROLOG: some 'working' illustrations 187

Appendix 2 'Probability' and 'lookout' programs 188

Appendix 3 Addresses of suppliers of software development tools 195

Bibliography 199

Index 204

INTRODUCTION

> THINK!
> - Could you ever consider a machine to be 'intelligent'?
> - If so, what would it have to be able to do to qualify?
> - If not, how would you define 'intelligence' to rule out machines?

Human beings have created many impressive tools which greatly surpass many of our natural abilities. With them we can travel great distances at great speed, lift huge masses, destroy on a vast scale. Yet because a tool can do certain things better than we can, we don't call it 'intelligent'. A car isn't intelligent because it can travel faster and further than we can run. A crane isn't intelligent because its ability to lift surpasses our own.

What, however, about tools which emulate and surpass our ability to perform *mental* rather than purely physical activities? Computers represent the most successful and widespread devices in this category. They can handle vast quantities of numerical and verbal symbols with impressive speed and accuracy, and few of us would disagree that they are extremely useful tools. Yet increasingly we can find statements from highly qualified individuals saying not only that machines can be intelligent but even alive! Some would even go as far as Geoff Simons, Chief Editor at the National Computing Centre (1983, p. 195), who considers that:

> There is overwhelming evidence that we are now witnessing the birth of a new family of living species on Earth—and this must be seen as one of the momentous events in the history of life.

At both extreme and more moderate levels there are many voices raised in the argument for and against the acceptance of the notion of intelligence in machines. This issue will be discussed in more detail in the final part of this book. For the moment what is really interesting is that the question of whether we can think of computers as intelligent, even alive, can and has been raised and sustained at a serious level. What this state of affairs means is that computers are beginning to emulate and surpass certain aspects of human *thought*; so much so that we must urgently turn our attention to a close examination of the nature, potential, even dangers of machine intelligence. In particular, those of us who are not computer specialists must nevertheless be able to achieve an informed understanding of what is being developed, and to add our voices to practical and philosophical aspects of the debate concerning how it can and should be applied. This book is aimed at the non-computer-specialist.

The range of mental activities which the computer can perform for us has greatly increased over recent years. Some of these are trivial or no more than routine in terms of what humans can do, although we have tended to associate them exclusively with human capabilities. Robots, for example, can learn a path through a maze or move objects around in seemingly intelligent ways. Other computer systems can converse with people, understand and write simple stories.

However, computers are also increasingly able to perform a range of what we think of as highly difficult and specialized intellectual tasks. Not only that—they are becoming as skilled and sometimes more skilled at them than their human counterparts.

So-called 'expert' computer systems already have an impressive list of achievements. We have examples of systems being used to great social and financial gain, as noted by Hayes-Roth *et al.* (1983, p. 6):

> PROSPECTOR has discovered a molybdenum deposit whose ultimate value will probably exceed $100 000 000 ...
>
> R1 configures customer requests for VAX computer systems ... despite the fact that even the resident experts thought it could not be done ...
>
> DENDRAL, which years ago demonstrated its superhuman performance, supports hundreds of international users daily in chemical structure elucidation ...
>
> CADUCEUS embodies more knowledge of internal medicine than any human and can correctly diagnose complex test cases that stymie human experts ...
>
> PUFF integrated knowledge of pulmonary function disease with a ... system for diagnostic consultations and now provides expert analyses at a California medical center.

High levels of performance have been achieved in other fields. Doug Lenat's AM (Automated Mathematician) system, for example, was designed to make mathematical discoveries. Soon after it started running it 'rediscovered' on its own many basic mathematical concepts and theorems, including Euclid's

Unique Factorization Theorem. It also discovered a number of interesting ideas unknown to its creator. Computer systems have now actually *created* new knowledge previously unattainable by human beings. Examples range from chess end-games to the diagnosis of heart-disease.

Interestingly, one of the most successful ways of devising 'intelligent' expert systems is not to design esoteric, highly complex computer procedures. Rather, it is to try to reproduce as accurately as possible in a computer system the relevant knowledge of human experts, gleaned from talking to and observing them. In this way, computers can not only 'think' as well as or better than humans, they can increasingly 'think' *like* humans.

For this reason, *human* thought will form both a starting- and a continual reference-point in this book in relation to the description and explanation of artificial intelligence—or the science of *machine* thought.

This book is designed for the non-computer specialist. Although details of programs are given (using the programming language PROLOG), it is possible to read these programs in plain English. English and PROLOG versions are given generally side by side. The main purpose of the book is not to teach you PROLOG (although you can, if you wish, pick up a very useful introduction to the language from it). Rather, it is to introduce many key ideas and principles used in work on artificial intelligence.

It tries to go some way towards solving a difficult problem in learning about artificial intelligence. You can often read about what sophisticated computer systems can do without any clear idea of *how* they do it. We hear of computers 'thinking', 'learning', using 'lines of reasoning', 'solving problems', and 'giving expert advice', but at a more basic level how can they do this? It's easy to see that computers can make calculations and print things on paper or on a screen—but in what way can they 'think'? To answer this, we must look at the details of programming. At this more detailed level it's often all too easy to get involved in learning about a programming language in great detail but be unable to see the wood for the trees. In other words, the sort of level at which you can understand programming is far removed from that at which sophisticated 'intelligent' computer systems operate.

There is bound to be a gap between the two levels, but this book aims to minimize that gap by building up an understanding of complex systems by first selectively looking at the details of programming. The examples of systems in Part C illustrate many of the concepts and ideas introduced at this more basic level in the earlier parts. This is made easier by the choice of programming language, PROLOG. This stands for PROgramming in LOGic, and PROLOG is much more closely related to how we think about problems than are the more usual languages such as BASIC, PASCAL, etc. In many important ways we can forget about the problems and intricacies of telling the computer exactly what to do at each stage of a program. Instead, we can concentrate much more directly on the nature of the task or problem the program is meant to be dealing with. For this reason, the level at which the non-computer expert can understand the programming details, and the level at which complex, 'intelligent' computer systems work, are much closer together. For this reason, this book includes both levels. Hopefully, you will be able to see a fair amount of both the wood and the trees making up the wood!

PROLOG is the language which the Japanese have chosen to form the basis of their ambitious program to develop fifth-generation computers capable of highly intelligent behaviour. It is already an important artificial-intelligence programming language, particularly in Europe, and is likely to increase in importance. There is also growing interest in teaching PROLOG in schools.

There are a number of different versions of PROLOG available for both large and small computers, including a number of home micros. The version used in this book is the one which is easiest to read, a version developed at Edinburgh University. It is the nearest we have to a 'core' PROLOG— a universally accepted common form. Other versions differ mainly in the punctuation and layout, and, to a large extent, one can be easily translated to another.

This book includes parallel English versions of the programs, so if you are following the book in PROLOG as well as plain English, and have access to a different version of PROLOG, you can easily translate into your own version.

Part A:

BASIC PROCESSES

CHAPTER 1

STORING KNOWLEDGE

A good starting-point in any search for comparisons and contrasts between human and machine thought is the question of *memory*. Man's collective memory—both in folklore and in libraries of recorded knowledge—is central to the development of his race. It allows each generation to function effectively on the basis of shared experience. It also allows each generation to rise increasingly higher on the shoulders of previous generations. In the same way, the memory of each person is vital to his or her ability to function and develop over time.

Memory is also of prime importance in the functioning and development of machine intelligence. Indeed, ways of representing knowledge in computer memory is a key area of research in artificial intelligence. Human beings can certainly represent large amounts of knowledge in their memories and use it flexibly and effectively. This is the basis for intelligent and creative thought and behaviour. Their weakness lies in their ability to forget and tire. Knowledge may often be partially or inaccurately remembered and applied. Efficient reasoning may not always be consistent.

Computers have complementary strengths and weaknesses. They can remember vast quantities of information and apply reasoning procedures accurately and tirelessly. However, their ability to use knowledge flexibly and creatively is (at present) more limited. Perhaps the true potential of artificial intelligence will lie not so much in the development of intelligent machines which will replace human expertise but rather in the linking of human and machine strengths and the minimizing of the weaknesses of each.

Let us first examine the nature of human memory. How do we store knowledge and how do we use it?

A common method employed by researchers to represent the contents of human memory is the *semantic network*. This consists of *nodes* (the dots in Figure 1) and *links* (the arrowed lines). A number of models of human memory have been developed which use semantic networks to represent knowledge. The close links between computer science and psychological research is

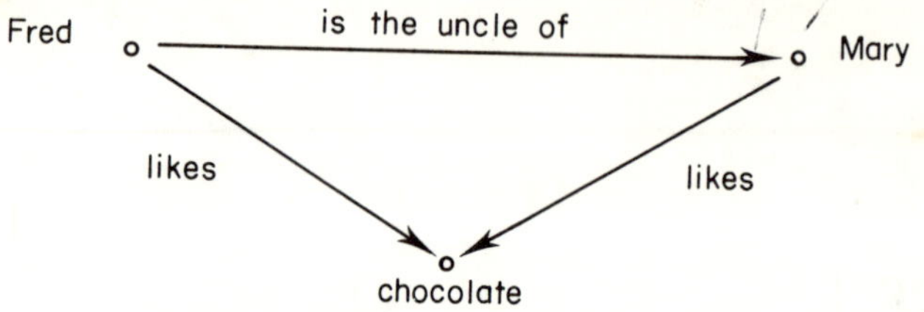

Figure 1. A semantic network

brought home by the fact that the first model of semantic memory—that of Ross Quillian—was constructed as part of a computer program, the 'Teachable Language Comprehender'. This was designed to be able to understand and produce language. The psychological validity of Quillian's model was subsequently tested and further models developed in the light of such studies. Two such models of memory are Anderson and Bower's 'Human Associative Memory', and Lindsay, Norman, and Rumelhart's 'MEMOD'. The models are very different in many ways but all use a system of semantic networks. Examples of these networks are shown in Figure 2. *Don't worry about the details.* The point is to illustrate different ways of using semantic networks made up of nodes and links. Much more complex ideas can also be represented in basically the same way, as illustrated in Figure 3.

THINK!

- Translate the semantic network shown in Figure 4 into seven short sentences. Here is one: 'John likes Anne'—you try the other six.
- Now arrange the sentences in the following pattern:
  verb ('subject', 'object').
  For example:
  likes ('John', 'Anne').

Now turn the page upside-down to see my answer!

The semantic network you have just translated could be typed into a PROLOG system in the following form:

PROLOG	ENGLISH
likes('Mary','John').	Mary likes John
likes('John','Anne').	John likes Anne
likes('Anne','Fred').	Anne likes Fred
dislikes('John','Mary').	John dislikes Mary
dislikes('Anne','John').	Anne dislikes John
envies('Mary','Anne').	Mary envies Anne
envies('John','Fred').	John envies Fred

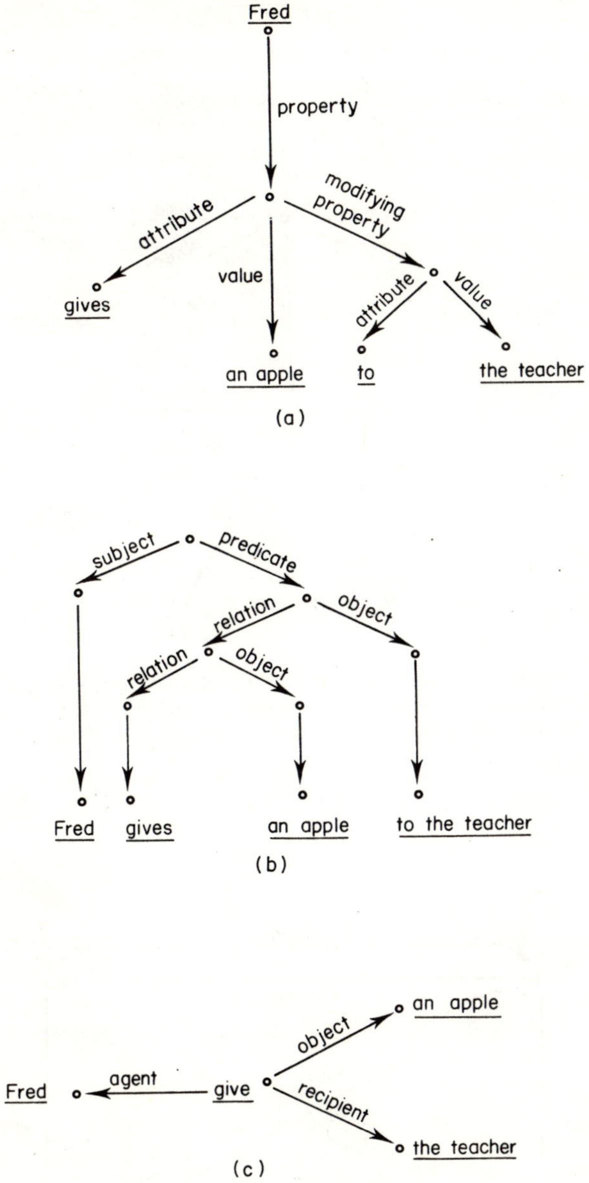

Figure 2. Models of how the sentence 'Fred gives an apple to the teacher' is stored in human memory. (a) Quillian's model: (b) Anderson and Bower's model; (c) Lindsay, Norman, and Rumelhart's model. (Adapted from Open University (1978). D303, *Cognitive Psychology,* Block 3, Units 18/19, p. 60)

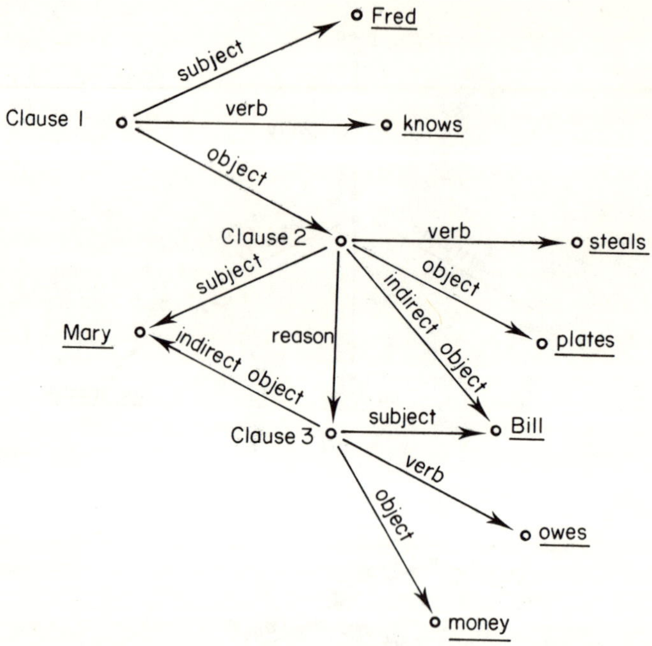

Figure 3. Semantic network representation of the sentence 'Fred knows that Mary steals plates from Bill because Bill owes Mary money'. (Adapted from Open University (1978). D303, *Cognitive Psychology*, Block 3, Units 18/19, p. 105)

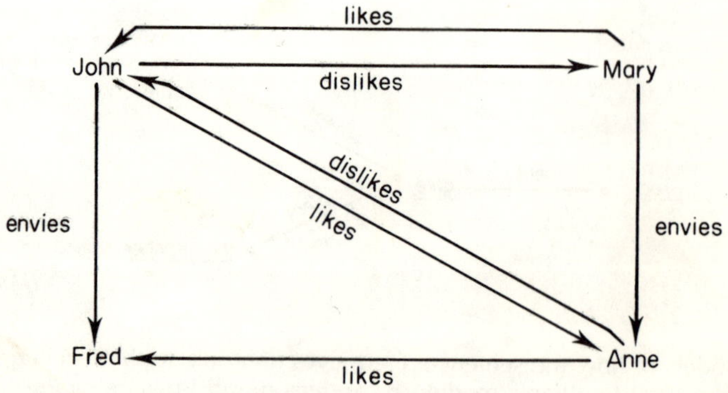

Figure 4. Another semantic network

You have just translated a semantic network into computerized form! As you can see, the switch from semantic network to lines of words, and vice versa, presents no problems, and it is this easy switching that represents the key to understanding how computers can handle knowledge in this form.

For those readers who have some familiarity with a computing language such as BASIC it will be interesting to note that there is no need to tell the computer that all this information is 'DATA' or to instruct it to 'READ' the information. There is no need, indeed, to write any program to tell the computer how to retrieve the information.

A number of programming languages now allow us to store and use knowledge of this type very easily. Ultimately all computer languages are able to do more or less the same things. As Amsterdam (1983, p. 150–1) has noted:

> It can be demonstrated mathematically that any computer language can simulate any other. So in terms of the problems they can solve, all computer languages are equivalent.

The differences lie in the ease with which different languages allow us to express certain types of problems and in the efficiency of the programs we can write. The particular language illustrated here is PROLOG. This is a relatively new language, growing in popularity, and particularly useful for work in artificial intelligence. It is interesting to note Amsterdam's comments (p. 148) that:

> ... the Japanese have chosen PROLOG to be the standard language for their fifth-generation computer project, an ambitious ten-year plan that, if it succeeds even partially, will make Japan the new world leader in advanced computer applications and expert systems. If the project succeeds, we'll all be programming in PROLOG by 1995.

Already built into PROLOG's system is an ability to answer logical questions. Type in, for example:

PROLOG	ENGLISH
?– likes('Mary','John').	Does Mary like John?

and PROLOG will reply:

Yes

Type in:

PROLOG	ENGLISH
?– likes('John','Mary').	Does John like Mary?

and PROLOG will correctly reply

No

We can also very easily ask questions like 'Who likes John?' by typing in

?– likes(X,'John').

In other words, we are saying 'Is there anyone—call him or her (X) for now because we do not know who it may be—who likes John?'. PROLOG will reply:

X = Mary

Once we have attuned ourselves to this rather roundabout way of asking questions we can ask more complicated ones. We can ask, for example, 'Is anyone *disliked* by someone whom he or she *likes*?' by typing in:

?– likes(X,Y), dislikes(Y,X).

↖— *the comma here means 'and'*

In other words, 'Does anyone (X) like anyone else (Y), and this second person (Y) dislikes the first person (X)?' In this case, PROLOG would reply:

X = Mary
Y = John

and go on to tell us who else is in this unfortunate position:

X = John
Y = Anne

Note that X and Y are what is known as *variables*. These are like refillable containers. In the example above, X could stand for *both* Mary *and* John, and PROLOG told us so by saying that:

X = Mary

and

X = John

In the version of PROLOG that we are using here *any* capital letter, or any word beginning with a capital letter, *if not enclosed by quote marks* is a variable. In this way, PROLOG automatically knows that X doesn't mean someone actually called 'X'. Rather, X stands for someone (or something) else—in this case both Mary and John. The fact that the real names Mary and John are enclosed by quotes tells PROLOG that these are real names and not variables.

THINK!	• How would you ask PROLOG the question: 'Which person is disliked by Anne, and envies someone liked by Anne?'

Here—upside-down—is the question translated into PROLOG form:

PROLOG	ENGLISH
?– envies(X,Y),	Is there a (Y) envied by an (X),
likes('Anne',Y),	and
dislikes('Anne',X).	Anne likes (Y) and
	Anne dislikes (X)?

When asked this question, PROLOG would reply:

X = John
Y = Fred

These examples are simple ones, but the structure and subject content could be as complex as you like. For example, statements may be 'embedded' within each other. In this way, the sentence:

'John knows that Mary likes Fred'

may be translated into PROLOG in the form:

knows('John',likes('Mary','Fred')).

PROLOG can then be asked what it is that John knows:

PROLOG	ENGLISH
?– knows('John',X).	What (call it X) does John know?

and reply:

X = likes('Mary','Fred')	What John knows (in other words, X) is that Mary likes Fred

Alternatively, the more specific question of who it is that John knows that Mary likes may be asked in the following way, again using X as the unknown quantity:

?– knows('John',likes('Mary',X)).	Who is it (call them X) that John knows that Mary likes?

to which PROLOG will dutifully reply:

X = Fred	The person (in other words, X) is Fred

The knowledge being handled in this way could equally well relate to electrical engineering or nuclear physics rather than the love-life of John, Fred, and Mary!

Programs may also be written to make PROLOG communicate in rather more normal-looking English! For the moment what is interesting to note is that the communication we have seen so far is built into the system. In other words, PROLOG will behave in this way without the user having to do anything other than type in the original facts—for example, likes('John','Mary').

CHAPTER 2

NEW KNOWLEDGE FOR OLD

So far, so good—but PROLOG is just giving us back what we have put in. It is searching through its memory, looking for a pattern of words which matches the pattern of words we are using to ask a question. We ask:

PROLOG	ENGLISH
?– likes('John','Mary').	John likes Mary?

and the system finds a statement in its memory which matches exactly (except for the question-mark) and replies:

Yes

Had no matching pattern been found, it would have replied:

No

When we add a 'variable', such as X or Y, for example:

?– likes('John',X).	Who (X) does John like?

the system interprets this as a request to tell us what the X replaces in the pattern, and replies:

X = Mary	The person (X) is Mary

In other words, PROLOG only 'knows' what you tell it—nothing different, nothing more. If we ask whether John is a man:

?– is_a('John',man).	Is John a man?

or even a person:

?– is_a('John',person).	Is John a person?

unless we actually *tell* PROLOG that John is a man, and a person by typing in is_a('John',man). is_a('John',person). PROLOG will not find any matching pattern, and will reply:

No

Hardly intelligent behaviour!

> THINK!
> - Imagine that you read in a newspaper that a new mammal—the 'quirn'—has just been discovered
> - Since you know that the quirn is a mammal, you know that *it can move, has blood, and suckles its young*. Yet the newspaper report did not specifically mention any of these things
> - Now write down the knowledge that you drew upon in your memory in order to know these things

This is an example of our interpreting new information in the light of our existing knowledge. My own version of the knowledge I used to infer what I was not explicitly told (that a quirn can move, has blood, and suckles its young) is shown below:

'If a quirn is a mammal, and all mammals
can move, have blood, and suckle their young,
then a quirn must *also* be able to move,
have blood, and suckle its young.'

This type of knowledge may be thought of as consisting of *rules* which allow us to go beyond the information explicitly given—to come to reasoned conclusions about something.

It is one thing to store descriptive facts, such as:

Mammals can move
Mammals have blood
Mammals suckle their young
A 'quirn' is a mammal
'Quirns' have just been discovered

It takes us into another dimension if we can also store rules which allow us to *use* these facts in an intelligent way. Such rules may be very specific. For example:

If a quirn is a mammal, and all mammals can move
Then a quirn can move

or more general. For example:

If animal (X) is a type of (Y), *and* all (Y)s can do
  a particular thing (Z)
Then animal (X) can also do thing (Z)

Rules can very easily be translated into computerized form. Using PROLOG, the simplest way of typing in the rules above would be as follows. (Note that the ':–' sign means *if* and the comma means *and*.)

PROLOG	ENGLISH
can(quirn,move):– is_a(quirn,mammal), can(mammal,move).	A quirn can move *if* a quirn is a mammal *and* a mammal can move
can(X,Y):– is_a(X,Z), can(Z,Y).	Something (X) can (Y) *if* (X) is a (Z) *and* a (Z) can (Y)

Once the latter rule had been typed in, PROLOG could use it to infer knowledge of new animals on the basis of what it knew about the species in general.

Although the examples here are simple, a computer system would be just as happy handling knowledge about more complex and serious matters:

If a patient is sweating and has a high temperature	*then* the patient has a fever
If the patient has a fever and a cough	*then* the patient has the 'flu
If the patient has a rash and a fever	*then* the patient has chicken-pox.
if the patient has the 'flu	*then* go to bed and take aspirin
if the patient has chicken-pox	*then* keep away from other people and go to bed

Not a particularly sophisticated level of medical diagnosis! However, it's interesting to note that basically the same type of rules—called 'production rules'—are used in complex 'expert systems' such as MYCIN, which gives diagnoses relating to bacterial infections in the blood, and PUFF, which identifies pulmonary disorders.

When the computer is told that the conditions on the left-hand side of a rule are satisfied (for example, that the patient *is* sweating and *does* have a high temperature) then it 'triggers' the right-hand part of the rule and asserts that the patient, in this case, has a fever. This piece of information can now be used where it is needed—in the second rule—to help satisfy conditions of other rules which, in turn, are 'triggered'. In this way, the system can use rules to infer what may be the matter with the patient.

We looked earlier at *semantic networks* as a way of representing human knowledge. These 'IF . . . THEN . . .' rules (or *production* rules) represent another technique. Semantic networks have been found useful both by psychologists interested in studying humans and by scientists developing computer systems. The same is true for production rules. As well as being a very useful and popular way of instructing computers they can be very helpful in studying human knowledge—in particular, knowledge of how to do things, rather than just facts and figures.

Some computer systems—such as ACT, which we will be looking at in Chapter 12—use both production rules and semantic networks working in close interaction.

THINK!
- Imagine that you are a doctor using the rules just described to find out what is wrong with a patient.
- How would your thinking differ according to whether you applied the rules from left to right (as just described) or, alternatively, from right to left?

Humans—and computers—can apply rules such as these either way round, that is, from left to right or from right to left (or a combination of the two). These different approaches are reflected in the varying ways of solving a particular problem.

Applying these rules from left to right, the doctor would be thinking:

'Let me ask the patient about his symptoms. On the basis of what he tells me I will decide what is the most likely diagnosis.'

A right-to-left approach, on the other hand, would imply a different reasoning process:

'I think that the patient may have measles. I'll find out whether he does in fact have the symptoms that I would expect if this guess is correct.'

In other words, the first approach implies an objective examination of all the evidence before any conclusion is drawn: the second represents the testing of a hunch about what the illness may be. In practice, either or both approaches may be used. We may gather one or two facts, which narrow down the range of possible conclusions. At this stage we may make one or two tentative guesses as to likely conclusions, and begin to check each one by asking very specific questions.

Computer jargon for these different approaches to using rules is *forward*- and *backward-chaining*. Forward-chaining refers to the first process—emphasizing evidence-gathering: backward-chaining refers to the second—emphasizing the testing of hypotheses. A number of computer systems, including SAVOIR (a system we will meet in Chapter 12) use both procedures in interaction. Forward-chaining is often referred to as 'data-driven' or 'bottom-up' processing; backward-chaining as 'goal-driven' or 'top-down'. These themes are also important in the study of human information-processing, and we will return to them in a subsequent chapter.

These approaches may be useful in solving certain types of problem—but human beings don't just solve problems then forget about them. Rather, we learn from experience. In other words, when faced with a problem which is the same as or similar to a problem we have previously met we draw upon our experience in order to solve the new problem more quickly or effectively. In this way we avoid having to solve from scratch the same sorts of problem each time they arise.

If the patient in our example were diagnosed as definitely suffering from chicken-pox then this information ought to be stored for future reference. Along with, or even instead of, details relating to all the specific symptoms reported to the doctor which caused him or her to reach this particular diagnosis some record would be made that the patient had chicken-pox. In other words, someone else looking at this person's medical record would not have to rediscover the diagnosis for him- or herself from the original symptoms.

In a similar way, it is easy to instruct a computer system to update—add to, change, and delete—its own memory. As well as diagnosing that patient (X) had chicken-pox the system could be told to add that fact to memory and, once the diagnosis were completed, to delete information about specific symptoms. In this way the computer would be changing its own knowledge store in the light of inferences it had made.

An amusing illustration of this ability would be to type the following information into PROLOG:

PROLOG	ENGLISH
kisses('John','Mary').	John kisses Mary
kisses('Mary','George').	Mary kisses George
kisses('George','Jane').	George kisses Jane
kisses('Jane','Fred').	Jane kisses Fred

then to add the rule that if someone (X) has a cold and kisses someone else (B), then this second person (B) will catch a cold too:

PROLOG	ENGLISH
catches(B,a_cold):–	(B) catches a cold if
has(X,a_cold),	(X) has a cold and
kisses(X,B),	(X) kisses (B),

We can go on to add on the last line an instruction to add to the computer's memory the fact that (B) now has a cold:

asserta(has(B,a_cold)).	Assert that (B) has a cold

'assert' is a PROLOG instruction to add something to memory—in other words, add it to its existing knowledge base. 'Asserta' means add it to the *beginning* of the knowledge base: 'assertz' means add it to the *end* of the knowledge base.

CHAPTER 3

SPREADING DISEASES AND TRACING RELATIONS—RECURSION

In real life, when someone catches a cold from someone else not only do they catch a cold they also 'catch' the ability to pass it on to other people. In other words, when someone infects you with a cold he or she also infects you with the power to infect. In this way, infections can spread throughout a group or society of people who come into contact with each other, in a sort of chain-reaction. The first person can be better long before a person further along the 'chain' has even caught it!

Building in a similar 'self-perpetuating' process is a very useful technique in not only human thinking but also computer programming. It is not confined to spreading diseases! It represents a powerful method of making inferences more generally. However, our analogy of spreading colds is a particularly useful way of illustrating the idea.

One way of writing an 'infection' program would be as follows:

PROLOG	ENGLISH
infect(A,B):–	(A) infects (B) if
has(A,a_cold),	(A) has a cold and
kisses(A,B),	(A) kisses (B) and
asserta(has(B,a_cold)).	add to memory the fact that (B) now has a cold

This would succeed in infecting one person—the one kissed by person (A). However, it doesn't give person (B) the power to go on and infect whoever he or she kisses. In order to obey the 'infect' rule, the computer will search its memory to find someone who has a cold to fill the variable (A). It will then find out who this person (A) has kissed, and duly note in its memory that this new person (B) now has a cold. Then it will stop—it won't bother trying to trace whom person (B) kisses unless we tell it to do so.

A particularly convenient way of telling the computer to carry on such a 'chain-reaction', automatically spreading the infection from person (A) to person (B), from (B) to (C), from (C) to (D)—as many times as necessary until it runs out of people who have been kissed by infected persons!—is to use something called *recursion*.

Recursion is rather like a message which includes instructions to make an exact *copy* of the message itself (including the instructions) and to pass the copy on to someone else. The basic idea is shown in Figure 5.

> Do you want to take part in the street party being planned for next Friday?
> Please write your name here ....
> and indicate whether you want to YES/NO.
> Return this form to John Smith, BUT BEFORE YOU DO—please make an exact copy of this note as you received it, and pass it to your next-door neigbour.
> J. F. Smith

Figure 5. A note using recursion

Translated into the 'infect' program, recursion could be achieved by including a simple instruction in the program to repeatedly apply *itself*—only with different people filling the (A) and (B) variables. This instruction is included on the last line:

PROLOG	ENGLISH
infect(A,B):–	Person (A) infects person (B) if
has(A,a_cold),	(A) has a cold and
kisses(A,B),	(A) kisses (B) and
asserta(has(B,a_cold)),	(B) now has a cold and
infect(B,C).	(B) goes on to infect someone else (C)

The last line reads: 'Go through the *infect* program again from the top, but as you do, "tip out" whoever you've got in the variables (B) and (C) in the *bottom* line *into* the (A) and (B) of the top line, and start again.'

Each time the computer cycles through the 'infect' program the people referred to as (A) and (B) will be *different*. Don't forget, (A) and (B) are variables—refillable containers.

When PROLOG applies a rule it tries to fill the first variable (refillable container) it comes to in the 'if' part of the rule (that part coming after the ':–'). In this case, this is the variable (A) in line 2. This particular line says: has(A,a_cold). So PROLOG looks for someone of whom this is true. It finds in its memory the sentence: has('Fred',a_cold). So it fills the variable (A) with *'Fred'*. It does so not just in line 2 of the rule but everywhere that (A) occurs in the rule. The rule now looks like this:

PROLOG	ENGLISH
infect('Fred',B):–	Fred infects someone (B) if
has('Fred',a_cold),	Fred has a cold and
kisses('Fred',B),	Fred kisses person (B) and
asserta(has(B,a_cold)),	(B) now has a cold and
infect(B,C).	(B) goes on to infect someone else (C)

PROLOG tries to fill in the *next* variable—in this case (B) in line 3. So it looks for someone who is kissed by Fred. It finds the sentence: kisses('Fred','Mary'). So it fills the variable (B) with *'Mary'*. The rule now looks like this:

PROLOG	ENGLISH
infect('Fred','Mary'):–	Fred infects Mary if
has('Fred',a_cold),	Fred has a cold and
kisses('Fred','Mary'),	Fred kisses Mary and
asserta(has('Mary',a_cold)),	Mary now has a cold and
infect('Mary',C).	Mary goes on to infect someone else (C)

The last line says: 'Go through the whole *infect* rule again, only this time put *Mary* in *Fred*'s place.' (We don't have anyone in (C) yet—but the old (C) of the bottom line will become the new (B) of the top line of the new cycle.) So the whole *infect* rule is applied again. This time it looks like this:

PROLOG	ENGLISH
infect('Mary',B):–	Mary infects someone (B) if
*has('Mary',a_cold),	Mary has a cold and
kisses('Mary',B),	Mary kisses someone else (B) and
asserta(has(B,a_cold)),	(B) now has a cold and
infect(B,C).	(B) goes on to infect someone else (C)

*(*Note*: Mary *does* now have a cold—line 4 of the rule added this fact the first time the rule was used above.)

Let's follow PROLOG through as it applies this program. First, we'll give it the following information, so that it will have something to work on:

PROLOG	ENGLISH
has('Fred',a_cold).	Fred has a cold
kisses('Fred','Mary').	Fred kisses Mary
kisses('Mary','John').	Mary kisses John
kisses('John','Jane').	John kisses Jane

Now we give it the rule:

infect(A,B):–	(A) infects (B) if
has(A,a_cold),	(A) has a cold and
kisses(A,B),	(A) kisses (B) and
asserta(has(B,a_cold)),nl,	add to memory that (B) now has a cold and go onto a new line (nl) and
write(B),write(' is now infected'),	write on the screen the name of person (B) followed by the words 'is now infected', and
infect(B,C).	(B) infects someone else (C)

The last line will make PROLOG repeat the whole program again and again, refilling variables (A), (B), and (C) with different people each time. The next-to-last line is there just to let you know what is happening as the program runs. Each time someone is infected, the person's name—stored in the variable (B)—will be printed out on the screen, followed by the words 'is now infected'.

Figure 6 shows the complete program, with names filling the variables at each stage as the program runs. We can start it off by typing:

```
?– infect(A,B).
```

PROLOG will reply:

```
Mary is now infected
John is now infected
Jane is now infected
No
```

After writing the names of all the people who are infected, PROLOG ends by writing:

```
No
```

This is because in fact the 'infect' rule fails. PROLOG keeps on searching for new people to infect until it runs out of people. At this stage, since it cannot

PROLOG

```
First stage      infect('Fred',Mary):-
                       has('Fred',a_cold),
                       kisses('Fred','Mary'),
                       asserta(has('Mary',a_cold)), nl,
                       write('Mary'),write(' is now infected'),
                       infect('Mary',C).
                              ↓     ↓
Second stage           infect('Mary','John'):-
                             has('Mary',a_cold),
                             kisses('Mary','John'),
                             asserta(has('John',a_cold)), nl,
                             write('John'),write(' is now infected'),
                             infect('John',C).
                                    ↓     ↓
Third stage                  infect('John','Jane'):-
                                   has('John',a_cold),
                                   kisses('John','Jane'),
                                   asserta(has('Jane',a_cold)), nl,
                                   write('Jane'),write(' is now infected'),
                                   infect('Jane',C).

                                   etc.
```

ENGLISH

```
First stage      Fred infects Mary if
                       Fred has a cold and
                       Fred kisses Mary and
                       add to memory that Mary now has a cold and on a new
                       line write on the screen that Mary is now infected, and
                       Mary infects someone else
                        ↓               ↓
Second stage           Mary infects John if
                             Mary has a cold and
                             Mary kisses John and
                             add to memory that John now has a cold and on a
                             new line write on the screen that John is now
                             infected, and
                             John infects someone else
                              ↓              ↓
Third stage                  John infects Jane if
                                   John has a cold and
                                   John kisses Jane and
                                   add to memory that Jane now has a cold and
                                   on a new line write on the screen that Jane is
                                   now infected, and
                                   Jane infects someone else

                                   etc.
```

Figure 6. The working 'infect' program, showing the names filling each variable at each stage of the program find anyone else, it thinks it has failed! We can easily get round this by including a second 'infect' rule:

```
infect(A,B).
```

Now when it runs out of people to infect, PROLOG will look for any other 'infect' rule which it can try instead of the first one (which has failed). It would find this second rule and apply it. This rule succeeds without doing anything! It does *not* say 'A infects B *if* something'—there is no 'if' part to the rule. It ends with a full stop. In other words, it simply says 'A infects B'. It is important to put this second 'infect' rule *after* the first one, so that PROLOG will only try it after it has tried the first. If it came to this rule first, the rule would succeed immediately and PROLOG would reply:

```
Yes
```

to our original question, without adding anything to memory or writing out anyone's name. If we put it afterwards, PROLOG will write:

```
Mary is now infected
John is now infected
Jane is now infected
Yes
```

No big deal to make PROLOG end by saying 'Yes' instead of 'No', but the difference is important. 'Yes' means that PROLOG has *succeeded* in satisfying the goal we set it when we asked it to:

```
?– infect(A,B).
```

In more complex programs, whether PROLOG succeeds or fails when trying to satisfy particular goals greatly affects the way the program runs.

However, recursion can be made much more effective and powerful than in this example. Keeping in the area of medicine, look at the imaginary 'medical family tree' shown in Figure 7. Undoubtedly a very healthy family!

Here we have a few generations of parents and children. The whole tree can easily be represented in PROLOG form as a series of sentences:

PROLOG	ENGLISH
child('Mary','Fred').	Mary is the child of Fred
had('Fred',flu).	Fred had flu
had('Fred',colds).	Fred had colds
child('Mary','Jane').	Mary is the child of Jane
had('Jane',colds).	Jane had colds
had('Jane',german_measles).	Jane had German measles
had('Jane',tonsilitis).	Jane had tonsilitis

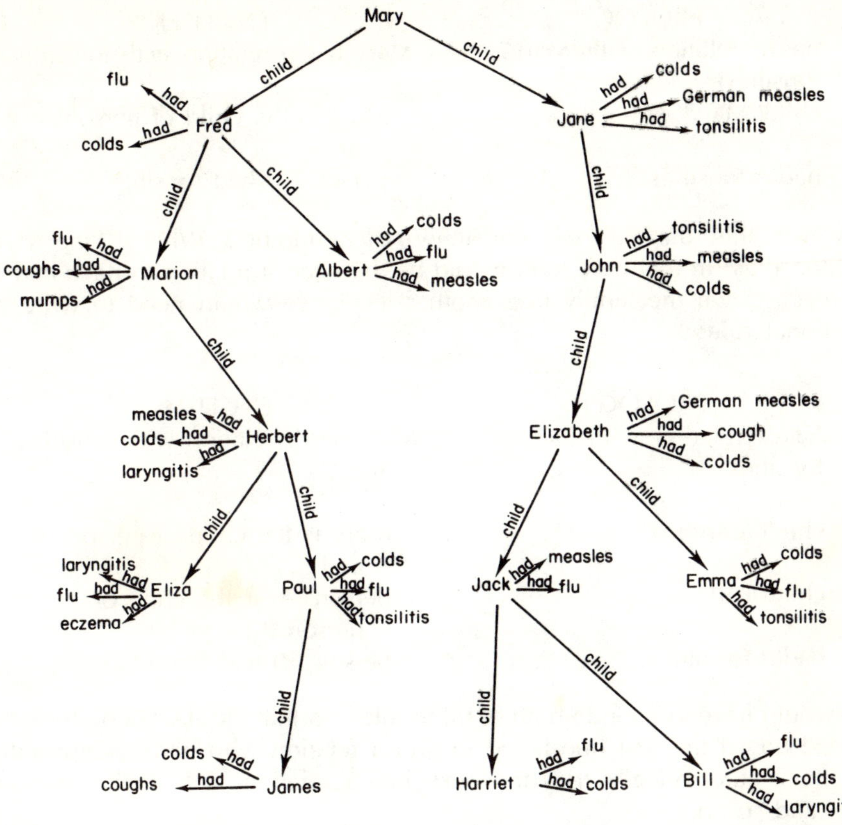

Figure 7. A medical 'family tree'

```
child('Fred','Marion').            Fred is the child of Marion
   had('Marion',flu).              Marion had flu
   had('Marion',coughs).           Marion had coughs
   had('Marion',mumps).            Marion had mumps

child('Fred','Albert').            Fred is the child of Albert
   had('Albert',colds).            Albert had colds
   had('Albert',flu).              Albert had flu
   had('Albert',measles).          Albert had measles

child('Jane','John').              Jane is the child of John
   had('John',tonsilitis).         John had tonsilitis
   etc.                               etc.
```

Now suppose we want to ask the question 'Has anyone in Mary's family ever had tonsilitis?' We can write a little rule *not* using recursion as follows:

PROLOG	ENGLISH
has_a_relation_with('Mary', tonsilitis):–	Mary has a relation with tonsilitis if
child('Mary',A),	Mary is the child of person(A) and
had(A,tonsilitis).	person (A) had tonsilitis

However, this rule will only find Jane (Mary's mother). What if Jane never had tonsilitis? In that case John would be the nearest relation to have had it. To reach down the family tree another level, we would need to write an additional rule:

PROLOG	ENGLISH
has_a_relation_with ('Mary', tonsilitis):–	Mary has a relation with tonsilitis if
child('Mary',A),	Mary is the child of person (A) and
child(A,B),	person (A) is the child of person (B) and
had(B,tonsilitis).	person (B) had tonsilitis

We would have to include both of these rules, so that PROLOG could try the second one if the first one failed to find a relation who had had tonsilitis. PROLOG automatically tries the next 'has_a_relation_with' if the first one fails, and so on.

The problem is that we would need a separate rule for each level of the medical family tree! For example, a rule to find relations at the next level down would be:

PROLOG	ENGLISH
has_a_relation_with('Mary', tonsilitis):–	Mary has a relation with tonsilitis if
child('Mary',A),	Mary is the child of (A) and
child(A,B),	(A) is the child of (B) and
child(B,C),	(B) is the child of (C) and
had(C,tonsilitis).	(C) had tonsilitis

That makes three rules, and we are only three levels down the family tree. *Recursion* allows us to achieve the same results with much less fuss. We can write two short rules which will find relations with tonsilitis, no matter how many levels down the tree they may be. As the tree grows more levels, the same rules will still work perfectly well.

THINK!

- It is possible to have only two rules which, when taken together, will find a relation with tonsilitis no matter how many levels the medical family tree has.
- I'll give you (in plain English rather than PROLOG) the first rule and part of the second rule. You try to finish it off!
- *Rule 1*: Mary has a relation with tonsilitis if Mary is the child of someone who had tonsilitis.
- *Rule 2*: Mary has a relation with tonsilitis also if Mary is the child of someone who ....
- Remember—no matter how large and complex the family tree, you will never need more than these two rules. The answer is given below.

PROLOG	ENGLISH
Rule 1:	
has_a_relation_with('Mary', tonsilitis):–	Mary has a relation with tonsilitis if
child('Mary',A),	Mary is the child of person (A) and
had(A,tonsilitis).	(A) had tonsilitis
Rule 2:	
has_a_relation_with('Mary', tonsilitis):–	Mary has a relation with tonsilitis if
child('Mary',A),	Mary is the child of person (A) and
has_a_relation_with(A,tonsilitis).	(A) *has a relation with* tonsilitis

These rules are *recursive*—in order to find 'someone who has a relation with tonsilitis' in the last line both the rules are applied again from the top. So in order to find the answer to part of the rules, the rules themselves are applied! Remember, though, that the name of the person filling the variable (or refillable container) will be different each time the rules are used in this recursive way.

If you find this tricky—don't worry. Even experienced programmers find recursion very difficult if they have not met it before! Perhaps it will be easier if we have a look at the program working in detail.

Look again at the family tree in Figure 7. Imagine that we want to ask PROLOG whether Mary has any relations who have had *coughs*. Note that neither of her parents have—the first person in the tree to have been troubled by them is Marion, Mary's grandmother on her father's side of the family. If we re-write the program substituting 'coughs' for 'tonsilitis', we can ask PROLOG:

PROLOG	ENGLISH
?– has_a_relation_with('Mary', coughs).	Has Mary any relation with coughs?

PROLOG will try to find a 'match' in its memory for this statement. If it can find a perfect match, it will reply:

Yes

if not:

No

Now here are the two rules showing how PROLOG applies them in this case. Remember that PROLOG will automatically try to apply the top rule first. If it doesn't succeed, it will go on to try the next one:

First rule

has_a_relation_with('Mary',	Mary has a relation with coughs
coughs):–	if
child('Mary',A),	Mary is the child of (A) and
had(A,coughs).	(A) had coughs

However, PROLOG cannot find anyone to fill the (A) slot—someone who had coughs *and* of whom Mary is the child. So this first rule fails, and PROLOG automatically goes on to try the next 'has_a_relation_with' rule:

Second rule

has_a_relation_with('Mary',	Mary has a relation with coughs
coughs):–	if
child('Mary',A),	Mary is the child of (A) and
has_a_relation_with(A,coughs).	(A) has a relation with coughs

First rule again

has_a_relation_with('Fred',	Fred has a relation with coughs if
coughs):–	
child('Fred','Marion'),	Fred is the child of Marion and
had('Marion',coughs).	Marion had coughs

The *second* rule says that Mary has a relation with coughs if Mary is the child not of someone who had coughs him- or herself but of someone who, in their turn, had a relation with coughs. In other words, in the second rule, PROLOG is trying to fill the (A) variable with the name of someone of whom Mary is the child *and* who has a relation with coughs. To find out whether anyone *can* be found who fits both categories the 'has_a_relation_with' rules are applied again in a second round—not this time to Mary but to a person of whom Mary is the child (person (A) in rule 2). Since the first relevant fact PROLOG will find is child('Mary','Fred'). Fred will be tried first.

In this second 'round' of applying the rules once again the first, top rule is tried first. Fred has a relation with coughs (Marion). A perfect match has been found, and PROLOG replies:

Yes

PROLOG is particularly powerful in this way because it can keep track of many 'rounds' or cycles of recursion. So it has no problems with very complex searches of the type:

Is there anyone who is the child of someone who
is the child of someone who
is the child of someone who
is the child of someone who
is the child of someone who
is the child of someone who
is the child of someone who
had coughs?

As they stand at the moment, the two rules are very limited. They can only be used to find a relation of one person—Mary—with one disease—coughs. We can make the rules much more general—able to find relations of any person, with any disease we specify—fairly easily. All we need to do is replace 'Mary' and 'coughs' with variables (A) and (B) as follows:

PROLOG	ENGLISH
has_a_relation_with(A,B):–	Person (A) has a relation with disease (B) if
child(A,C),	(A) is the child of person (C) and
had(C,B).	(C) had disease (B)
has_a_relation_with(A,B):–	Person (A) has a relation with disease (B) if

child(A,C),	(A) is the child of person (C) and
has_a_relation_with(C,B).	(C) has a relation with disease (B)

We can now ask PROLOG:

?- has_a_relation_with('Fred', laryngitis).	Has Fred a relation with laryngitis?

and PROLOG would reply 'Yes'. Or we could ask:

?- has_a_relation_with('Jane', german_measles).	Has Jane a relation with German measles?

and PROLOG would reply 'Yes'.

Note that the program is still very limited. The only people it can trace are those linked by parental relationships. For a 'working' illustration of this recursive program, have a look at Program A in Appendix 1.

Very complex rules can be applied recursively, giving PROLOG the power to solve extremely complex problems. A slightly more complex type of use is shown in the next chapter.

CHAPTER 4

DO IT YOURSELF

Based on what you know so far, let's build a very simple 'expert system' designed to identify a number of animals when told certain characteristic features.

We'll built the system at three levels of sophistication. This is so that we can see how the techniques of recursion (which we have just met) and list-handling (which we haven't yet) can be used to advantage.

Version 1

The first version of the animal-identification system is very unsophisticated but it does provide the foundations for understanding the more advanced versions we'll be meeting soon. It makes no use of recursion or list-handling. We'll see how these techniques can be used to increase the versatility and presentation of our system in versions 2 and 3.

This small system will be able to recognize a set number of birds and mammals. These are:

ostrich, penguin, giraffe, tiger, cheetah

Now the first thing we must decide is what knowledge our system is going to need if it is to recognize each particular creature. This knowledge can take the form of a number of features characteristic of each creature. My set of features which will identify an ostrich is, in plain English and PROLOG:

PROLOG	ENGLISH
features(ostrich,if(has_feathers, has_long_neck,does_not_fly)).	Features of an ostrich are—has feathers, has a long neck, doesn't fly

> THINK! • Now you try making some rules for the other creatures.
> • Remember—try to make them as succinct as you can: i.e. try not to include characteristics that lots of animals have. It would be a waste to say that, for example, you can tell a tiger because it has skin, blood, etc.!

They probably won't be identical, but, hopefully, there should be some common ground. Here are mine:

features(ostrich,if(has_feathers, has_long_neck,does_not_fly)).	Features of an ostrich are—has feathers, and a long neck, doesn't fly
features(penguin,if(has_feathers, swims,does_not_fly, is_black_and_white)).	Features of a penguin are—has feathers, doesn't fly, is black and white
features(giraffe,if(has_long_neck, has_long_legs,eats_leaves)).	Features of a giraffe are—has a long neck, has long legs, eats leaves
features(tiger,if(has_black_stripes, is_orange,eats_meat, has_four_legs)).	Features of a tiger are—has black stripes,is orange, eats meat, has four legs
features(cheetah,if(has_dark_spots, runs_very_fast,eats_meat, has_four_legs)).	Features of a cheetah are—has dark spots, runs very fast, eats meat, has four legs

We now need a program which can make use of this knowledge:

identify:–	You will have completed the 'identify' program *if* (in other words *when*) you have
features(X,Y),	Found a 'features' list and filled the (X) variable with the name of a creature, and the (*Y*) with its list of characteristics. (PROLOG will automatically start with the topmost 'features' list—*ostrich*)
nl,write('Tell me (yes. or no.)'),	written on a new line (nl) the

	words 'Tell me (yes. or no.)' on the screen,
nl, write(Y), nl,	Followed on a new line (nl) by whatever is in (Y)—in the first case this will be 'if(has_feathers,has_long_neck, does_not_fly)',
read(Z),	Stored whatever the user replies (yes. or no.) in a variable labelled (Z),
Z = yes,	Checked that this reply (in variable (Z)) is 'yes'. (If it *is*, then PROLOG goes on to the next line of this program. If it is *not*, then the whole of this program fails and PROLOG goes back to the beginning and starts again with the *second* 'features' list (i.e. *penguin*). If 'penguin' also fails, it will go on to try the third (i.e. giraffe), and so on.)
nl, write('The creature is a '),	Written on the screen 'The creature is a '
write(X).	Followed by whatever is in the (X) variable (in the first case, *ostrich*)
	If the 'identify' program above has failed for all the creatures listed, then PROLOG will automatically try this second 'identify' program, which simply says
identify:–	You will have completed the 'identify' program when
write(' I do not know this creature ').	You have written on the screen ' I do not know this creature '.

To start this program, the user would simply type:

?– identify.

The program would reply:

Tell me (yes. or no.)
if(has_feathers,has_long_neck,does_not_fly)

If the user then types:

yes.

The program would announce:

The creature is a ostrich

However, if the user were to type:

no.

then the program would reply:

Tell me (yes. or no.)
if(has_feathers,swims,does_not_fly,is_black_and_white)

If the user were to reply 'no.' to all the questions for all the creatures, the program would finally say:

I do not know this creature

Version 2

We can make this mini 'expert system' more sophisticated in a number of ways. First, we'll design it so that it can classify an animal as a mammal or bird, then as carnivore or vegetarian, and so on, as well as giving us the name of the actual animal itself. The knowledge which will form the basis of the system is the classification shown in Figure 8.

We can denote the features which allow us to recognize various animals as follows:

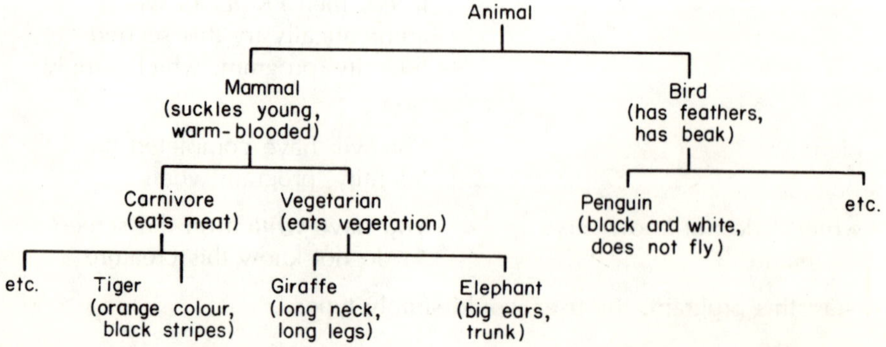

Figure 8. Knowledge-base for version 2 of the 'animal-identification' program

features(animal, mammal, if(suckles_young, warm_blooded)).	The features of the animal 'mammal' are—that it suckles its young and is warm-blooded
features(animal, bird, if(has_feathers, has_beak)).	The features of the animal 'bird' are—has feathers and has beak
features(mammal, carnivore, if(eats_meat)).	The features of the mammal 'carnivore' are—eats meat
features(mammal, vegetarian, if(eats_vegetation)).	The features of the mammal 'vegetarian' are—eats vegetation
etc.	etc.

Once again, we must now tell PROLOG what to *do* with this knowledge in order to recognize a particular animal. In the following program, which does just this, I have shown variables *not* by (X), (Y), etc. but instead by more meaningful labels (Name), (Subspecies), (Features), etc. In the version of PROLOG we are using here, *anything* beginning with a capital letter is a variable, unless it is enclosed by quotes. So (Characteristics) will be treated by PROLOG as a variable, ('Characteristics') would not:

PROLOG	ENGLISH
Recognize(Name):–	To recognize a creature (call it *Name* for the moment)
features(Name,Subspecies, Characteristics),	First find the 'features' list for the *Name* you've been asked about (the user will ask: ?– recognize(animal). to start the program. Thus the *Name* variable will first be filled with 'animal')
write('Tell me (yes. or no.)'),	Write on the screen the words: Tell me (yes. or no.)
nl, write(Characteristics),nl,	Followed on a new line by whatever is stored in the *Characteristics* variable. The first

	animal features list PROLOG will find is: features(animal,mammal,(. . .etc.) so the *Characteristics* variable will çontain if(suckles_young,warm_blooded)
read(Reply), Reply = yes, nl,	Read the user's reply, and check that this is in fact 'yes.'. If it is *not*,
	then the program fails and PROLOG will go back automatically to try the next 'features' list for the *Name* (in this case, *animal*) it has been asked to recognize. If the reply *is* yes., then PROLOG will carry on to the next line -
write('The creature is a '),	In the case of the first *animal* features list—for mammal—these lines will tell
write(Subspecies), nl,	PROLOG to write on the screen:
	The creature is a mammal

THINK!

- So far, the program will recognize an animal if it is a mammal or a bird—but no more. In other words, it will ask whether the animal suckles its young and is warm-blooded. If it *does*, the program will say:

  The creature is a mammal

- However, we want it to go on and find out what type of mammal the creature is (carnivore or vegetarian); and then what type of carnivore or vegetarian (tiger, giraffe, etc.).
- One more line of program will achieve this. Thinking back to the examples of *recursion* in the last chapter, can you make up a line which will cause the program to go on to recognize the subspecies of mammal, and on to recognize the subspecies of carnivore or vegetarian, and so on, no matter how many subdivisions there are? One line of program will make the program recursive, so that it goes on to recognize as many subdivisions as there are.

The last line of the program should read:

recognize(Subspecies).

This causes the program to begin again at the top, only substituting for the original *Name* (filled originally by 'animal') whatever is now in *Subspecies*. If the animal has been identified as a 'mammal', the new contents of *Name* (emptied out from the present contents of *Subspecies*) will be 'mammal'. So PROLOG now tries to recognize 'mammal', and so on, recursively.

To use the system, the user would type:

?– recognize(animal).

The system would reply:

Tell me (yes. or no.)
if(suckles_young,warm_blooded)

If the user types:

yes.

The system would respond:

The creature is a mammal

Tell me (yes. or no.)
if(eats_meat)

User:

yes.

System:

The creature is a carnivore
Tell me(yes. or no.)
if(orange_colour,black_stripes)

User:

yes.

System:

The creature is a tiger

The system will work perfectly well as it is but untidily! It will end by failing, as PROLOG tries to find what subspecies there are of tiger. We don't have any subspecies of tiger in our knowledge-base, so the system would fail. It would also fail if it could not recognize a particular creature, rather than saying 'Sorry, I don't know this animal'. We can easily remedy this situation by adding the following lines:

recognize(tiger).	When you try to apply the program to 'tiger', you have finished your work—there are no more subdivisions
recognize(giraffe).	Same when you try to apply it to giraffe
etc. to cover all the animals listed in the knowledge-base.	Same for all the animals listed in the knowledge-base
recognize(_):–	If the previous 'recognize' rules have all failed, i.e. you haven't identified one of the animals listed above and haven't been told to stop, then
write('Sorry, I don't know this animal').	Print on the screen the words 'Sorry, I don't know this animal' (The underline sign on its own is an 'anonymous variable': basically in this position it means 'anything', and so would work for any animal, including one *not* in the knowledge-base)

It is important that the last 'recognize (_):–' rule comes *after* all the others so that PROLOG only applies it as a last resort.

Although this system is small it does illustrate some of the important features used in more complex, large-scale expert systems. The same sort of procedure can be applied to very large and complex problems.

Version 3

We can improve version 2 in a number of ways. First, we can save the user the effort and annoyance of having to answer the same questions twice. This didn't occur in versions 1 and 2, but imagine a more complex knowledge-base, in which the same fact could support several different conclusions, depending on which other facts it is combined with. For example, having a long neck could be a distinctive feature of a giraffe if it were combined with having four legs and eating leaves. When combined with laying eggs and not flying it would signify an ostrich! The point is that we don't want the system to ask whether the creature has a long neck when it is considering the possibility of its being a giraffe then ask the same question when considering whether it might be an ostrich. This would be tedious from the user's point of view. In complex knowledge-bases this sort of problem may be widespread—more so than in our animal example.

Let's extend our knowledge-base to include this sort of problem, as shown in Figure 9.

In order to make the improvements we are after we need to deal with each feature separately. In other words, we need to be able to ask the user whether the creature has whiskers and to note the answer, and whether it is tame and to note the answer, etc. as separate questions rather than as part of a 'package-deal'. This will avoid the user being asked, for example, 'Tell me if the creature *has whiskers*, and is tame', then 'Tell me if the creature *has whiskers*,

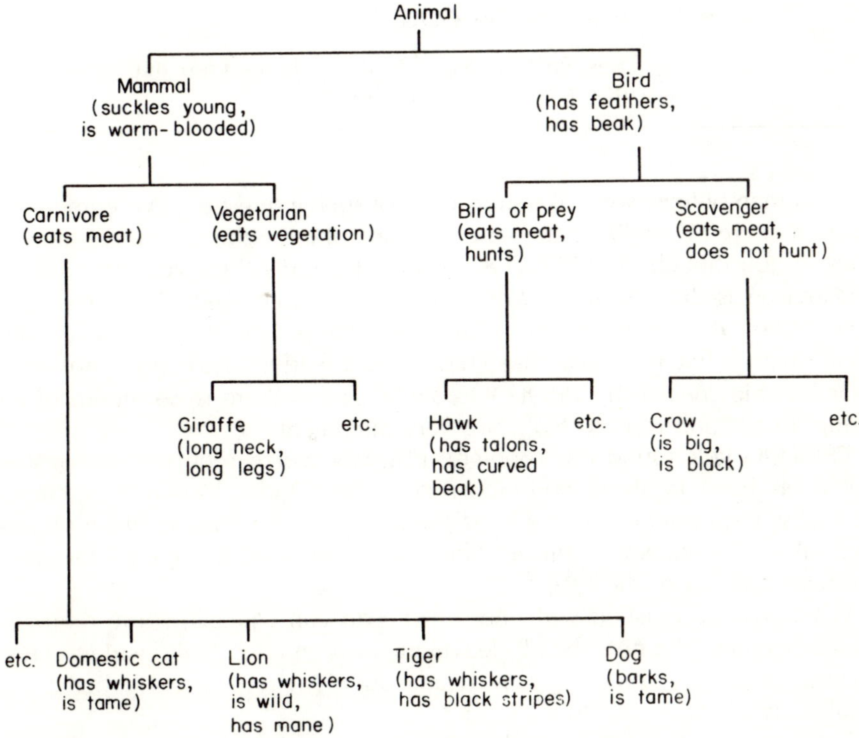

Figure 9. Knowledge-base for version 3 of the 'animal-identification' program

is wild, and has a mane', then 'Tell me if the creature *has whiskers*, and black stripes', etc.

To do this, we need to use *lists*. Using lists will allow us to 'get at' individual characteristics, no matter how many there are for any given creature. However, we need to know something about how PROLOG copes with lists.

Coping with lists is something human beings do entirely naturally. Unfortunately, computers find this much harder! The techniques we have looked at so far, including recursion, can all be made to do much more if they can handle them. Appendix 1 includes 'working' examples of lists but I would advise you to read the following explanations before looking at them in detail.

THINK!
- Imagine that we type into PROLOG the following facts:

PROLOG	ENGLISH
likes('Fred',chips,peas, fish).	Fred likes chips, peas, fish

- When we ask PROLOG:

?– likes('Fred',peas).	Does Fred like peas?

will PROLOG reply 'Yes' or 'No'?

As human beings, we make short work of such a question. We immediately know that Fred does like peas—they are one of three things that we know he likes. If you thought that PROLOG would also reply 'Yes' you are mistaken! The reason is that PROLOG knows one thing—that Fred likes 'chips,peas, fish'. When it is asked whether Fred likes 'peas', PROLOG tries to match 'peas' with what it is told that Fred likes. Clearly, 'peas' does not match exactly 'chips,peas,fish'. PROLOG concludes that there is no record of Fred liking 'peas', and replies 'No'. Not very intelligent!

PROLOG gets round this problem in a way which seems to us—who do not even have to think consciously how to handle lists—rather clumsy. Basically, it splits all lists into a *head* and a *tail*. Any list is rather like toothpaste in a tube! You squeeze a bit out (the head of the list), leaving all the rest in the tube (the tail of the list).

When you've finished with the 'head' you can squeeze some more out, which becomes the new 'head', leaving a (now smaller) 'tail' inside the tube. So you can go on, squeezing new 'heads' out of the tube, leaving increasingly smaller 'tails' inside the tube.

Just as you can see and use the bit of toothpaste you've just squeezed out (unlike the rest left in the tube) so PROLOG can process and use the head of a list much more than the tail. For example, if we want to count the number of items in a list:

cake, bread, fish, peas, chips

we must tell PROLOG to split the list into a head (cake) and a tail (bread, fish, peas, chips) and count the head (one item). Then we must tell it to break the tail into a new head (bread) and tail (fish, peas, chips), to count the new head (one item), and add it to the first head (making a total of two items). Then break the remaining tail into a new head (fish) and tail (peas, chips), count the head, and add it to the total (now three items), and so on.

If we want PROLOG to print each item of a list on a new line, we can tell it to break the list into a head (cake) and a tail (bread, fish, peas, chips)—to print the head then go onto a new line and then do the same with the tail. So the remaining tail is broken into a new head (bread) which is duly printed, and a tail, and so on.

THINK! • How would you make PROLOG find out whether an item was included in a list—for example

Is 'chips' a member of the list 'cake, bread, fish, peas, chips'?

To see if an item is included in a list we must do the same. We must ask PROLOG to break the list down repeatedly into a head and a tail, and each time compare the word it is given (chips) with the new head of the list. When the two match, PROLOG will reply:

Yes

PROLOG has a special way of showing lists—they appear in square brackets. The list:

chips, peas, fish

would be written in PROLOG as follows:

[chips,peas,fish]

The head of this list is *chips*, the tail is *peas,fish*. We can tell PROLOG to print the head of the list in the following way:

print_the_head([A|B]):–	You will have printed the head of a list whose head we'll call (A) and whose tail we'll call (B) when you have

write(A).	Written (= printed on the screen) whatever is in the variable (A)

We can refer to the head and tail separately if we call one (A) and the other (B)—these letters are just variables as in previous chapters—and split them by a |. When PROLOG finds the square brackets it knows it is dealing with a list. When it finds the | it knows that what comes in front of it is the head of the list, and what comes after it is the tail. In this case, the head (A) is in fact *chips*. The tail is all the rest—*peas,fish* which itself can be split into a head and tail.

To 'get at' all the items on the list (in other words, to print the whole list not just the first item) we can tell PROLOG to break the list into a head and tail, to print the head, go onto a new line, then break the tail into a new head and tail, print the new head, then do the same with the tail, and so on. We can do this as follows:

print_the_list([A|B]):–	You will have printed the list with head (A) and tail (B) when you have
write(A),nl,	Written the head (A), and gone onto a new line (nl)
print_the_list(B).	And then gone through the same program again (recursively) from the top, only this time applying it not to the original list but to the tail (B) of the original list (which is itself a list). The (B) of the bottom line will be split into a new head (A) and a new tail (B) in the top line of the program

This program is recursive—it can go on re-applying itself. Each time it does so, it breaks the tail of the previous 'round' into a fresh head and tail. It does this until it is left with no more items in the list. (Remember—what goes into the (A) and (B) variables will be different each time round.)

We can tell PROLOG when to stop (when there are no more items in the list) by saying:

print_the_list([]).	When you find an *empty* list to print, do nothing more— The full stop, instead of ':–' means you have finished

We can put this little rule anywhere. Assuming we put it before the other one, PROLOG will come to it first in each 'round'. Each time it comes to it whilst there are still items in the list it will fail (because the list is not empty so will not match the [] in the rule). So it will go on to the second rule. When the list *is* empty, the [] in the first rule *will* match, and PROLOG, on finding the full stop, will do no more with these particular rules. 'print_the_list' will have succeeded.

Both rules, then, will look like this:

print_the_list([]).	When you find an empty list, stop.
print_the_list([A|B]):–	To print a list with head (A) and tail (B)
write (A), nl,	You print the head (A), go onto a new line, and then
print_the_list(B).	Substitute the tail (B) for the original list and apply the whole program to it from the top

So much for lists in general—but how can we use them in our animal-identification program? We can store information about the characteristic features of each animal in lists, as follows:

features(animal,mammal, [suckles_young,is_warm_blooded]).	The animal 'mammal' suckles its young and is warm-blooded
features(animal,bird,[has_feathers, has_ beak]).	The animal 'bird' has feathers and has a beak
features(mammal,carnivore, [eats_meat]).	The mammal 'carnivore' eats meat
features(mammal,vegetarian, [eats_vegetation]).	The mammal 'vegetarian' eats vegetation
features(bird,bird_of_prey, [eats_meat, hunts]).	The bird 'bird of prey' eats meat and hunts
features(bird,scavenger,[eats_meat, does_not_hunt]).	The bird 'scavenger' eats meat and does not hunt
features(carnivore,domestic_cat, [has_whiskers,is_tame]).	The carnivore 'domestic cat' has whiskers and is tame
features(carnivore,lion,[has_whiskers, is_wild,has_mane]).	The carnivore 'lion' has whiskers, is wild and has a mane
features(carnivore,tiger, [has_whiskers,has_black_stripes]).	The carnivore 'tiger' has whiskers and has black stripes
features(carnivore,dog,[barks, is_tame]).	The carnivore 'dog' barks and is tame
features(vegetarian,giraffe, [has_long_neck,has_long_legs]).	The vegetarian 'giraffe' has a long neck and long legs

PROLOG	ENGLISH
features(bird_of_prey,hawk, [has_talons,has_curved_beak)].	The bird of prey 'hawk' has talons and a curved beak
features(scavenger,crow[is_big, is_black]).	The scavenger 'crow' is big and black

An important thing to note about lists is that they can be of any length—so the number of features of different creatures need not be the same and the program will still work!

Now we must tell PROLOG what to *do* with this knowledge in order to recognize a particular animal. I have put boxes round each section of the program to make things clearer.

PROLOG	ENGLISH
recognize(domestic_cat). recognize(lion). recognize(tiger). recognize(dog). recognize(giraffe). recognize(hawk). recognize(crow).	When asked to recognize any of these animals—go no further. There are no further subdivisions of any of them
recognize(Name):–	To recognize a creature (call it *Name* for the moment)
features(Name,Subspecies,List), nl,nl,	First you find the 'features' rule for the *Name* you've been asked about (the user will ask '?– recognize(animal).' to start the program—so the *Name* variable will first be filled with *animal*) ('nl' means 'go onto a new line')
check(List),	Then find the 'check' rule (below) and apply it to what you found in the *List* variable above (i.e. the list of defining features)
write('The creature is a '),	Having followed all the instructions you found in the 'check' rule above, then write on the screen the words 'The creature is a '

write(Subspecies),nl,	followed by whatever is presently filling the *Subspecies* variable
recognize(Subspecies),nl.	Now go through the whole program again from the top, only apply it to whatever is at present in the *Subspecies* variable. In other words, whatever is in the *Subspecies* variable in the bottom line of the program will be 'tipped out' into the *Name* variable in the top line the second time through. And so on.
recognize(_):–	If you have run out of animals and all the other 'recognize' rules have failed, then
write('I do not know this animal').	Write on the screen 'I do not know this animal', then stop

check([]).	You arrive here from the 'check(List),' line of the 'recognize(Name):–'. If the list is empty, then do nothing and return to the program where you left it. 'check' has succeeded
check([A|B]):–	If the list is not empty, then break it into a head (A) and a tail (B)
ask_about(A),	Take whatever is in the head (A) and apply to it whatever instructions you find in the 'ask_about' rule (below), *then* go on to the line below this one
check(B).	Now put the tail of the list (B) through the same two 'check' rules (above), breaking it down into a new head (a new A), and a new tail (a new B)

ask_about(A):–	To ask about whatever is in (A)
fact(A,yes).	Find out whether the user has already been asked and has replied 'yes.'. If he or she *has*, there will be a fact (based on the user's reply) stored in memory. (See the third 'ask_about(A):–' rule to see how this is done. Before any questions have been asked, PROLOG will automatically apply this third rule, since the first two 'ask_about(A):–' rules will fail). If the user *has* replied 'yes.', then this rule succeeds. Carry on the program where you left off (i.e. the last line of the 'check([A|B]):–' rule). If *not* then go to the next rule below
ask_about (A):–	Alternatively, to ask about whatever is in (A)
fact(A,no),	Find out whether the user has already been asked and has replied 'no.' If this *is* the case, then
fail.	Fail. The next 'ask_about(A)' rule will be tried but will also fail. This failure will be handed back to the 'check' rule, which in turn will hand back the failure to the 'recognize' rule, and a new animal will be tried
ask_about(A):–	Come to this rule if the previous two rules have failed.
not(fact(A,_)),	Check that the user has *not*

	already been asked and has replied either 'yes' or 'no'. (The underline sign _ is a variable meaning 'anything'. In this case it can stand for both 'yes' and 'no'.) If the user has *not* already been asked, then PROLOG carries on to the next line
write('Tell me (yes. or no.) if it '),	Write on the screen 'Tell me (yes. or no.) if it ',
write(A),nl,	followed by whatever is stored in (A). This will be one of the features (e.g. 'suckles_young' or 'has_feathers')
read(Reply),	Store the user's reply ('yes' or 'no') in a variable called *Reply*
assertz(fact(A,Reply)),	Add to memory a new fact, using whatever is in (A) *and* the user's reply stored in *Reply*, e.g. fact(suckles_young,yes).
Reply = yes.	Check that the user's reply is in fact 'yes'. If it *is*, then carry on where you left off the program (i.e. the last line of the 'check' rule). If the user's reply is *not* 'yes', then this rule will fail. This failure will be handed back through the program. In effect, this means that PROLOG is given the message: 'The subspecies you had in mind and were asking about is *not* the animal in question. Try asking about the *next* subspecies'

THINK!

- I will give you the program lines which will make the system explain its line of reasoning.
- Your task is to decide *where* in the program they should go.
- The idea is that each time the system says:

  The creature is a mammal (or whatever else it is)

  It goes on to say:

  because suckles_young
  because is_warm_blooded

- Here are the lines (the first line will be located separately from the other five). Where do they go?

explain(List),	Explain the list of features
explain([]).	If the list of features is empty, the explanation is finished
explain([A|B]):–	If the list is not empty, break the list into a head (A) and a tail (B), and
write(' because '),	Write ' because '
write(A), nl,	Followed by the head (A), e.g. 'because suckles_young'
explain(B).	Then go on to do the same with the tail of the list (B)

The first line that I gave you should go after the fifth line of the 'recognize(Name)' rule—after 'write(Subspecies),nl,' and before 'recognize(Subspecies)'. The other lines could go anywhere. This is because, like all rules, they form a relatively 'self-contained' set of instructions which the system will look for under its name wherever it is put.

A final improvement is needed before the system can be used repeatedly. At the moment, each time a creature is identified, new facts are added to memory. So if we identify an ostrich and then ask the system to identify a second creature the features of the first creature will still be in the computer's memory and would cause confusion. Some animals would have truly amazing accumulations of characteristics!

So we need a final instruction to make the system, once it has recognized the animal in question and all its subspecies, clear its memory of that creature's characteristics. In this way, the system will be ready to recognize the next creature.

We can use the PROLOG command 'retract' here. 'retract' simply means 'delete from memory'. If we tell the system to:

retract(fact(has_whiskers,yes)).	Forget the fact that the creature has whiskers

this fact will be forgotten. However, we want the system to forget all the facts concerning the particular animal just recognized. To do this, we need a small set of rules which we could call 'clean_slate':

clean_slate:–	To clear the system of all facts
retract(fact(X,Y)),	Find the first fact, and fill (X) and (Y) with the details—(X) will be a characteristic, (Y) will be either 'yes.' or 'no.'. Delete from memory the fact that (X) and (Y)
fail.	Fail—in other words, go back and try to find another fact to retract. PROLOG will repeatedly find another fact, and retract it until all facts are retracted. Only when all facts are retracted in this way will PROLOG go on to the second 'clean_state' rule below—still convinced it has 'failed' in the first rule!
clean_slate.	PROLOG will only come to this rule when it has retracted all the facts it can find using the first 'clean_slate' rule (above). This second rule says 'Succeed by doing nothing' (the full stop). This goal therefore succeeds and the system's task is finished

The instruction to 'clean_slate' will be put as the last line of the 'recognize(Name)' rule—after the 'recognize(Subspecies)' line (which must now end with a comma instead of a full stop to allow PROLOG to go on to the next line. The final two lines of the 'recognize(Name)' rule will now read:

```
...
recognize(Subspecies),nl,
clean_slate.
```

We also need to put it at the end of the 'recognize(_)' rule, which will now read:

```
recognize(_):–
write('I do not know this animal'),
cleanslate.
```

The *rules* for 'clean_slate' telling PROLOG what to do when told to 'clean_slate':

```
clean_slate:–
... etc.
```

can go anywhere—it is simplest to put them at the end of the program. They will be searched for by name when the system is told to use them.

Part B:

STRATEGIES FOR INTELLIGENT BEHAVIOUR

CHAPTER 5

PUTTING YOUR (MATHEMATICALLY) BEST FOOT FORWARD

So far, we have looked at some basic processes involved in storing and using knowledge. This knowledge took the form of facts about the world and rules which allow new facts to be formed.

However, these basic processes on their own do not guarantee high levels of intelligent behaviour or 'high-performance thinking'. It is the same with any type of high performance. Something over and above basic processes is needed.

For example, how long would Steve Cram last on a tennis court with John McEnroe? How long would it take René Arnoux to lap them both on the race-track? Yet all three men have very similar 'basic processes'—arms, legs, brain, level of fitness, etc. All are highly intelligent and there is no reason to suppose that each has different physical circuitry or different basic reasoning processes. High performance relates to higher-level strategies: in other words, not to our basic processes but to how we organize and use them.

Indeed, the basic processes underlying highly intelligent behaviour can be used to produce relatively unintelligent behaviour—particularly where computers are involved!

Computers are far better than human beings at remembering and applying rules unfailingly and exhaustively. Given a set of symptoms shown by a patient, a computer could faithfully apply all its relevant 'if—then' rules and list all possible diseases likely to cause those symptoms. Working 'backwards' instead of 'forwards', for any given disease, it could work out possible symptoms and ask whether they were present in the case being diagnosed. Either way, it could compute all the possible results of applying its rules.

In many cases, however, such a 'brute-force' approach may not work. The number of possible combinations of symptoms may be too great to be useful and manageable. Also it is not characteristic of human experts simply to try to generate all possible permutations of symptoms or diseases by the exhaustive application of rules. An essential part of human expertise is rather to develop the ability to follow up the most profitable leads and select the most likely interpretations. The expert chess-player, to take another example, does not attempt to compute for each move all possible outcomes: rather he or she thinks in terms of broad plans, strategies, and patterns developed on the basis of experience, such as 'try to gain control of the centre of the chess-board'.

THINK!

- Imagine a very fast computer playing draughts. If it uses the 'brute-force' approach and works out for each possible move all the possible subsequent moves which would lead to a win, loss, or draw, how long would it take:
  (1) A minute?
  (2) An hour?
  (3) A week?
  (4) A month?
  (5) If none of the above, say how long.

Here is the answer:

Full marks if you thought I had underestimated slightly. In fact it has been estimated that it would take approximately a thousand million billion centuries, as there are ten thousand billion billion possible moves. The point is that for some tasks, using a purely brute-force approach will take a little too long!
(*Source*: Open University (1978). D303, *Cognitive Psychology*, Block 4, Units 26/27, p. 20)

This is known in the jargon as the problem of the 'combinatorial explosion'. Human beings are particularly good at coping with it and taking enlightened short-cuts in problem-solving.

The jargon word for these short-cuts is 'heuristics', and they are the key to high levels of intelligent behaviour. They are higher-level knowledge which guide, and keep to manageable proportions, the workings of lower-level knowledge.

Let's imagine once again a doctor trying to diagnose a patient. The patient complains of having a high temperature, sickness, vomiting and diarrhoea. What the doctor does *not* do is try a 'brute-force' approach and list all possible diseases all having these symptoms. Hopefully, he or she shows rather more intelligence.

> THINK!
> - Put yourself in the doctor's position. Assuming that you reject a 'brute-force' approach, what approach do you use instead?
> - Think what heuristics—short-cuts—you would take to diagnosing the illness. Note them down.

Probably what you would do would be to listen to the symptoms and build up a mental 'short-list' of the most likely causes. You could then quiz the patient further as you follow up these likely illnesses. The seriousness of certain illnesses could also influence your thought. For example, even if there was a fairly small chance of the illness being a particularly serious one, if some of the symptoms *did* fit it would probably be worth following up even if only to rule it out at an early stage. In other words, you would make intelligent decisions. You would not just work through every possibility in maximum detail in an unthinking way. If you did try to do this, you would just run out of time and energy.

There are various ways in which the process of cutting down the impossibly vast range of possible lines of thought and action involved in complex problem-solving can be built into computerized systems. Let's try, for example, to find the shortest route through a complex series of pathways—a concept which is surprisingly useful for solving a wide variety of problems.

Even playing a simple game may be thought of as finding a way through such a series of paths. Imagine the familiar toy: a plastic square containing numbers and one empty space to allow the numbers to be moved around one by one. The numbers start in jumbled order and the idea is to shuffle them around until they form a correct sequence. In Figure 10 I have shown the pathways associated with a (very) miniaturized version of this game—in fact I have used only three numbers in order to make the explanation clearer to follow.

In Figure 10 the top square represents the starting-point of the game—imagine that when we pick it up it is in this order, 2 and 3 on the top, with the 1 under the 3. In artificial-intelligence jargon this order is the *initial state*. The square at the bottom of the figure is how we want the order to end up—in jargon, the *goal state*.

The two squares underneath the initial state represent the only two possible moves which can be made from the initial state. These are either to move the

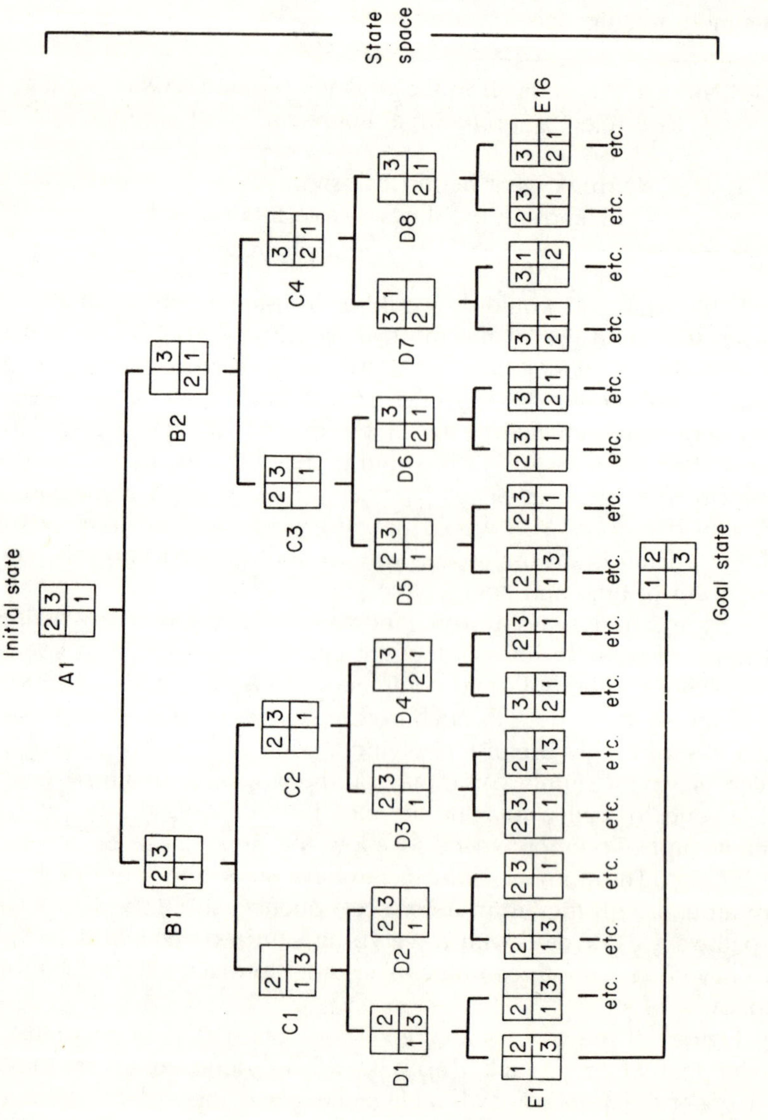

Figure 10. Possible pathways to a solution

1 across into the empty space or to move the 2 down into the space. Each of these possible moves leaves the way open for two next possible moves. If our first move had been to move the 1 across to the empty space then *from* that position we could make one of two further possible moves—either moving the 3 down into the empty space or moving the 1 across into the empty space (not a good move—this would bring us back to our starting-point, the initial state!), and so on.

The whole network of pathways—all possible moves that can be made—is called the *state space*. This space can be searched in a variety of ways, but before we look at possible ways note that the *goal* state (the solution) is reached in Figure 10 at the very bottom left-hand square. The shortest path through the state space, then—in other words, the quickest way of solving the problem—is, beginning at the initial state, to choose the move represented by the square at the *left* each time. (It is, of course, not usually so simple to find the best path in such problems. I have put it as the left choice in each case in order to make it easier to illustrate!)

A 'brute-force' approach to finding the best route through the network of pathways would be simply to try every possible route in order, finishing when you happened (by luck or exhaustively searching) to hit on the goal state. However, this may be the very last square you explore, if you are unlucky, and with state spaces vastly bigger than this one, as in any complex problem, this may be time-consuming or, in the case of very large state spaces, impossible.

A 'depth-first' brute-force search would involve trying out (on Figure 10) squares in the order A1, B1, C1, D1, E1. If this did not arrive at the goal state the next sequence to be tried would be to 'backtrack' and try E2, D2, E3, E4, C2, D3, E5, E6, etc.—in other words, going down the tree then backtracking to the next untried paths if unsuccessful.

A 'breadth-first' brute-force search would involve the sequence A1, B1, B2, C1, C2, C3, C4, D1, D2, D3, D4, D5, etc.

THINK!

- Can you think of a way of making more intelligent choices—in other words, a way to decide which of moves B1 or B2 you should make, then which subsequent move to make, and so on?
- Assume that you cannot see the whole state space. You know what goal state you are aiming for, but can only see, at the time of choosing a move, the two possible choices open to you.
- In other words, think of a way of avoiding the risk of having to try every path, by at each stage reducing your choice to the most promising.

One way of making more intelligent, as opposed to brute-force, choices is to give each alternative a *score*, depending on how nearer it brings you to the goal state (the solution), and to choose always the move with the highest score (bringing you closest to the goal state).

For the example shown in Figure 10 we could base a score on how far away each number is in the squares from where it is in the goal state—in other words, how far away it is from where we ultimately want it to be. We could, for each possible move (i.e. each square), add up the score for all the three numbers to make a score for the whole square. The bigger the score, the further away from the goal state. Therefore we would need to always choose as our next move the square with the lowest score. Let's see if that works.

Compare the two choices offered by squares B1 and B2. Which should we choose?

Goal state

1	2
	3

Square B1

2	3
1	

Square B2

	3
2	1

Looking first at square B1, the number 1 is only one place away from number 1's position in the goal state. So we'll give it a score of 1. Number 2 is also only one position away—score that 1 as well. Number 3 is also only one position away—score 1. So the total score for square B1 is *3*.

In square B2, number 1 is two positions away from where it needs to be (remember—you can't move diagonally within the squares!). Number 2 scores two as well. Number 3 scores one. So the total score for square B2 is *5*. That means we should choose square *B1* and not B2 as our next move.

Having chosen B1 we can do exactly the same to our next two possible moves—squares C1 and C2.

Goal state

1	2
	3

Square C1

2	
1	3

Square C2

2	3
	1

THINK! • Work out the scores for squares C1 and C2.

Square C1
Number 1 scores one; number 2 scores one; number 3 scores zero: the total score for square C1 is therefore two.

Square C2
Number 1 scores two; number 2 scores one; number 3 scores one: the total score for square C2 is therefore four.

So we choose move C1, and so on.

Similar sorts of procedures can be applied to less trivial problems. Figure 11 shows a road network. We want to go from City A to City E. It is obvious from the map that if we go from City A to City B we are further away from our destination than when we set off. The quickest route would seem to be A→C→E, and we could easily calculate this by scoring our choice of route according to how much nearer our choice would bring us to our destination, as in the previous number-game example.

However, in real life things are rarely so simple! Imagine, for example, that the road shown going from City B to City D is a motorway and all other roads are minor roads. In terms of finding the quickest route from A to E our simple numerical calculation wouldn't be much help. Our best route would undoubtedly be to go from City A to join the motorway as soon as possible at City B, to travel by motorway to City D, then leave it and drive to City E.

However, this would involve making a first choice, from A to B, which results in our being further away from our destination than when we started.

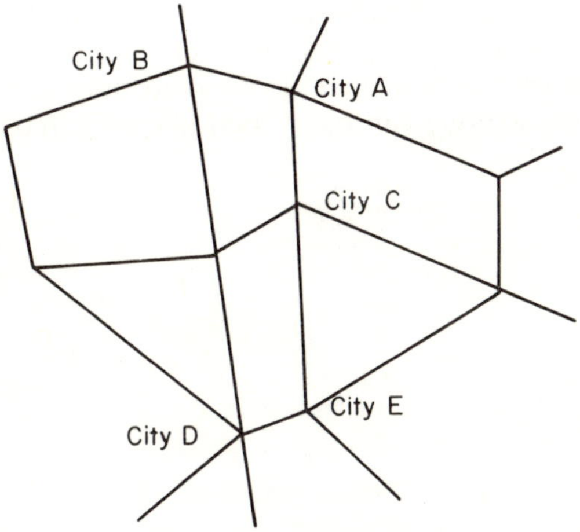

Figure 11. A road network

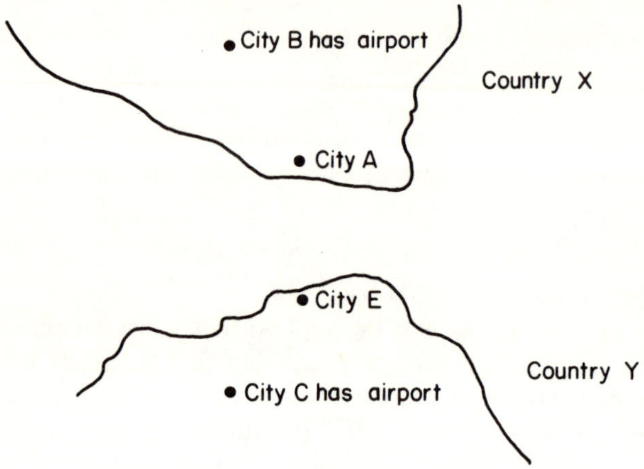

Figure 12. International travel

Clearly, we are using rather more sophisticated knowledge than simple calculation of distance from our destination.

A similar problem would arise if we wish to travel internationally. Imagine travelling from City A to City E in Figure 12, for example. The quickest route may well be to travel to the airport at B, fly to C, then travel by road E. On the other hand, if the roads are busy and there is a good sea-ferry service, the quickest route may be by sea directly from A to E. The point is that our simple numerical calculations of distance are less relevant than rather more sophisticated knowledge about how to get from place to place. Typically, planning a journey in real life involves using such knowledge to break down the planning into a series of subplans—for example, to go by road to the airport, to fly to another airport, then go by road from that airport to your destination. Let's see how we can build such planning into a computerized system.

CHAPTER 6

PLANS AND SUBPLANS

Imagine that we want to travel from Bolton, Lancashire, to Murcia, southern Spain. It would not take us long to formulate the plan:

> First go by bus from Bolton to Manchester Airport. Fly to Alicante Airport, then take a bus to Murcia.

This would be one plan. Someone else may prefer the train instead of the bus. However, basically, the plan would consist of the three subplans: (1) journey to Manchester; (2) journey to Alicante; and (3) journey to Murcia.

We can devise a computer program which will break down plans into subplans in a similar way. In other words, rather than simply setting off, hoping to get nearer and nearer a solution the program will break the problem down into component steps and try to solve each one in an appropriate way using more sophisticated knowledge.

Before looking at the program in detail, let's first get an overall idea of how it will work. Basically, it will solve our 'Bolton to Murcia' travel problem in the way shown in Figure 13. Have a good look at this, and follow the flowchart through in detail to get the idea clear.

The same program could solve similar travel problems involving other cities and countries (provided that details of them were in the computer's memory), and could solve much more complicated travel problems requiring many more subplans. The problem illustrated can be solved in three travel stages—bus to Manchester, aeroplane to Alicante, and bus to Murcia—but the same program could deal with large numbers of such stages. More importantly, very much more complex problems have been tackled and can be dealt with by using more complex versions of basically the same problem-solving method. This particular technique is known as *means–ends analysis.*

Because it is designed to be able to deal with different cities and different stages, the problem-solving parts of the program make no mention of Bolton,

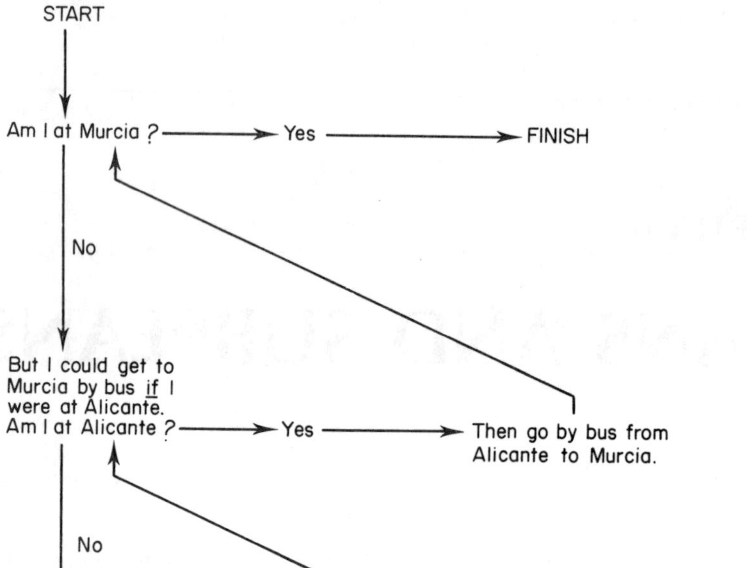

Figure 13. Basic idea behind a program to solve the 'Bolton to Murcia' problem

Alicante, or the like. They approach the problem more generally, and refer to place (Y) and place (Z). These labels are variables (refillable containers). The names of particular places can be put in, taken out, and replaced, as the problem requires. Bearing this in mind, and in conjunction with Figure 13, have a look now at the program—not yet in PROLOG-form but rather in diagram-Form, as shown in Figure 14. Try to tie it in with Figure 13.

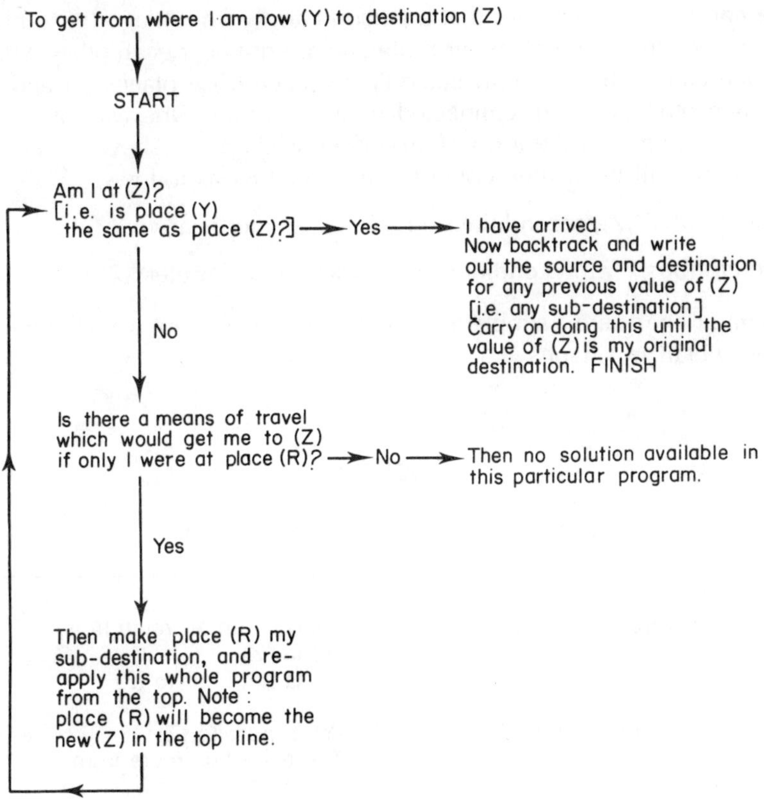

Figure 14. English version of a program to solve the 'Bolton to Murcia' problem

We would also need to give the computer a fair amount of general information including the following facts:

PROLOG	ENGLISH
air_route(manchester,alicante).	There is an air route from Manchester to Alicante
near_to(bolton,manchester).	Bolton is near to Manchester
near_to(alicante,murcia).	Alicante is near to Murcia
bus_route(bolton,manchester).	There is a bus route from Bolton to Manchester
bus_route(alicante,murcia).	There is a bus route from Alicante to Murcia

We can tell the computer that a person can fly from place (Y) to place (Z) *if* (Y) and (Z) are linked by an air route and are not near each other. Similarly, someone can go by bus from place (Y) to place (Z) *if* places (Y) and (Z) are near each other, and are connected by a bus route. One way of displaying this information is in the form of an 'operator chart', as shown in Figure 15.

This could all be written concisely in PROLOG as follows:

```
operator(fly,Y,Z,preconditions(air_route(Y,Z),not(near_to(YZ)))).

operator(bus,Y,Z,preconditions(near_to(Y,Z),bus_route(Y,Z))).
```

Operator charts are a simple form of 'frame'—a concept we will be meeting in more detail in Chapter 9.

ENGLISH

operator	*preconditions*
Fly from (Y) to (Z)	There is an air route from (Y) to (Z) (Y) is not near to (Z)
Bus from (Y) to (Z)	(Y) is near to (Z) There's a bus route from (Y) to (Z)

PROLOG

operator	*preconditions*
fly(Y,Z)	air_route(Y,Z) not(near_to(Y,Z))
bus(Y,Z)	near_to(Y,Z) bus_route(Y,Z)

Figure 15. An 'operator chart'. (Adapted from Open University (1978). D303, Cognitive Psychology, Block 4, Units 26/27, p. 44

So much for the information we have given the computer. Now here are the instructions on what to do with that information—the problem-solving rules. Although they may seem daunting at first, don't worry. Have a look at them and try to get some idea of what they are doing. I will explain them in more detail in a moment.

PROLOG	ENGLISH
travel(Z,Z).	If the place I'm wanting to travel to is the same as my present location, then I need do nothing at all!
travel(Y,Z):–	But if not, then to travel from place (Y) to place (Z)
operator(A,R,Z,preconditions(B,C)),	Find an operator (A) which will get me to (Z) from somewhere else (R)
call(B), call(C),	Check that preconditions (B) and (C) are satisfied (e.g. places (R) and (Z) do share an air route and are not near to one another)
travel(Y,R), nl,	Apply the 'travel' rules recursively from the top, but substitute place (R) for the original destination—in other words, make getting from my present location to (R) a sub-problem, or one stage in the journey. Whatever is in the variable (R) will become the new (Z). PROLOG will keep a record of which place name occupies the (Z) variable in any given cycle
write([A,' from ',R,' to ',Z]).	The 'travel(Y,R)' line above will only be done when the first 'travel(Z,Z)' line fires. This will only be when the program has cycled through to the point where my destination is where I am located.' The success of

the first 'travel(Z,Z)' rule passes back to the 'travel(Y,R)' rule which called it. PROLOG can now go on to the 'write' line, and write on the screen the values of the variables (A), (R) and (Z) as they exist in the last recursive cycle. PROLOG will then go back through any previous cycles (i.e. previous journey stages), and write the contents of the variables as they existed in each cycle. When PROLOG has returned through all the cycles to the original goal, then the whole program is successfully completed

Working with our example, let's substitute real places for the (Y) and (Z)—which would be the case if the program were actually being used. We'll keep it simple at first, and just try to get from Manchester to Alicante. We would start by asking PROLOG:

```
?– travel(manchester,alicante).
```

The first 'travel' rule would be tried, but would fail since both 'manchester' and 'alicante' cannot be (Z) at the same time! The first 'travel' rule cannot match with the query we have asked PROLOG.

So the second 'travel' rule is tried, and this time the variables *can* be filled ('instantiated') as follows:

travel(manchester,alicante):–	I can travel to Alicante from Manchester *if*
operator(fly,manchester,alicante, preconditions(air_route(manchester,alicante), not(near_to(manchester, alicante)))),	There is an operator which would get me there assuming that there is an air route between Manchester and Alicante, and that they are not near to one another
call(air_route(manchester,alicante)),	And if there *is* a statement in the database that there is an air route

call(not(near_to(manchester, alicante))),	and if there is *not* any statement that Manchester and Alicante are near one another
travel(manchester,manchester), nl,	and if I can travel from Manchester (my original starting point) to Manchester (my new sub-destination)

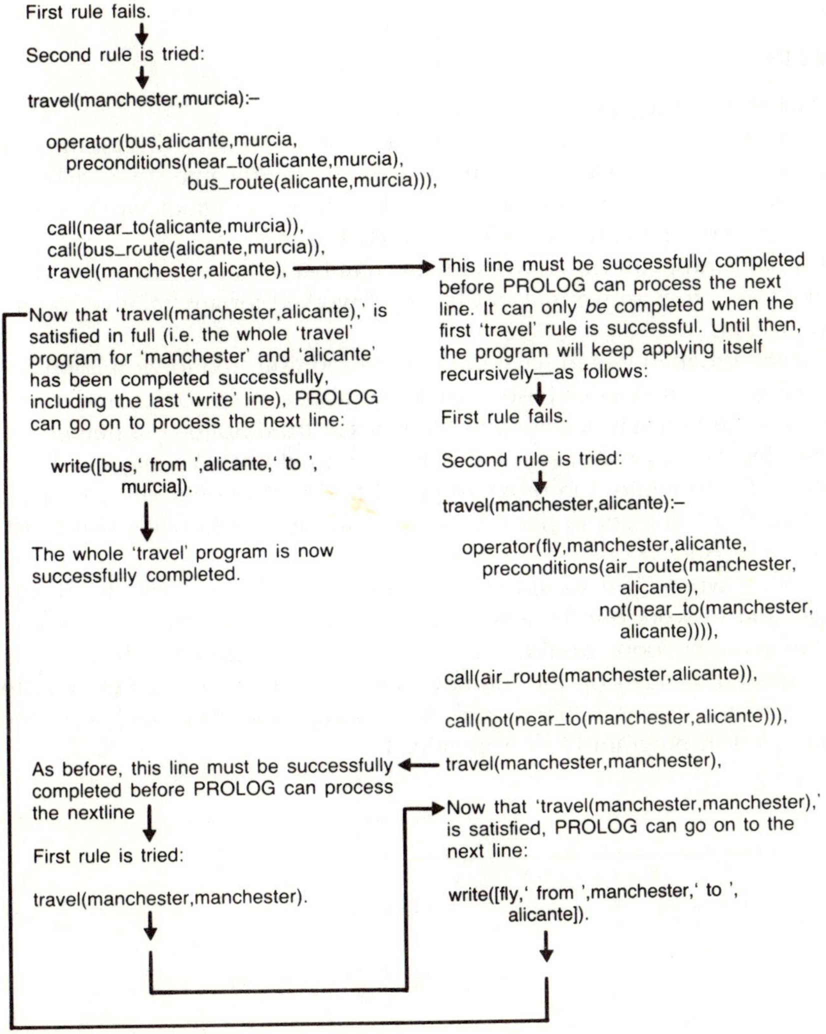

Figure 16. The travel program in action

The bottom line recursively calls the whole 'travel' program again from the top line. This time the first 'travel' rule succeeds, since the two (Z)s both refer to 'manchester'

```
travel(manchester,manchester).
```

This first rule is something of a 'dummy' stage in the journey since no travelling is needed. All it does is pass the message back that a successful route has been found. The line that called it—the penultimate line of the second 'travel' rule—is now satisfied, so PROLOG can go on to process the next (and last) line:

```
write([fly,' from ',manchester,' to ',alicante]).
```

But what if our journey is not quite so simple?

Imagine that we want to get from Manchester to *Murcia*. This journey would require two steps—a plane from Manchester to Alicante, then a bus from Alicante to Murica. Figure 16 shows the 'travel' program working on this example, with the variables filled in at each stage.

The same program would work for any number of stages, just as the 'medical family tree' program we met in Chapter 3 worked for any relative, no matter how distant.

But the program would have to be more sophisticated to work with a large and complex knowledge-base. If it knew about more towns and cities we may well find ourselves getting to Manchester from Bolton via numerous bus routes depending on which one the program tried first! We may even get into a loop—for example, the program could find that I could get from place A to place B if I first got to place C. I could in turn get to place C if I first got to place B. And so on.

In other words, we would need to add a number of rules to avoid any loops, and to work out the best choice of operators at any given time. One way of avoiding loops would be to keep a record of cities which have already been passed through on the journey, and to check that any new city to be tried does not appear in that record. A 'working' example of such a procedure is included in program D in Appendix 1.

CHAPTER 7

MORE SOPHISTICATED STRATEGIES

Seeing the wood before the trees

> THINK!
> - The approach we've been looking at breaks large problems down into smaller subproblems, but only when all details in all the subproblems have been worked out can the whole plan be put into action.
> - Can you think of the *limitations* of this approach—for example, when planning a holiday in real life?

This approach is terribly logical but if you or your friends carried it to its logical conclusion you might correctly be accused of very *un*intelligent thinking. You would in fact be trying to tie up every small detail before proceeding. Without sufficient cash in your pocket you couldn't pay the taxi-driver to your hotel. Without this precondition you couldn't book your flight, but you are unlikely to make the success of the whole plan hinge on making sure in advance that you would have the right change! These things are mere details and should not be allowed to have undue influence on more important matters. They can be solved nearer the time, when more important things have been seen to, or even on the spot when you arrive!

In the sort of programs described so far the computer had no idea of what were more important and less important factors. It couldn't tell that not having the right change for a meal on board the plane, or to pay the taxi-driver, although annoying, are less important—and more easily solved—than finding yourself booked on the wrong flight.

You would be even less likely to be worried about such details if you were still at the stage of browsing through holiday brochures, wondering where to go, and whether you would like to travel by air, sea, or road. In other words, much intelligent thinking involves making broad plans—where you want to go, how you want to get there, how long you can afford to stay—before worrying about details that can be sorted out once you've decided. In short, we need to be able to plan at a more general, abstract level and to avoid getting bogged down in details.

One way of building a similar capacity to this more general, abstract thinking into a computer is to give it priorities. As we have seen, operators like 'fly' and 'bus' have *preconditions* which must be satisfied before they can be applied. There were only three preconditions in the travel example given, but in more complex problems operators may have rather more. If we can tell the computer which are the most and least important preconditions then it can only deal in the first instance with very important ones. In this way, it can form a high-level, general plan taking account of only the most critical preconditions. Lower-level, less critical preconditions may well be unsolved at this level, but despite this, the computer can sketch a broad tentative plan. If this seems workable and acceptable, then a lower level of detail (in which lower-level preconditions are considered) can be worked on.

This can be done by giving the preconditions simple numerical 'criticality values'. Imagine that an operator 'go_by_taxi' has three preconditions that must be satisfied before it can be successfully applied—telephoning for the taxi, having enough money to pay the driver, and giving the driver directions.

Having enough money is probably the most critical precondition which we can give value 1—no use worrying about telephoning, and certainly no use thinking about giving the driver directions until you've sorted the money out! Telephoning for the taxi is probably next—give it value 2. The least critical precondition is giving the driver instructions—no need to worry about this until you are sitting in the taxi.

When working out a complex plan using many operators the computer could work out a grand master-plan by considering only the most critical preconditions of all the operators. If the master-plan seemed to work, it could check out the details at a lower level, then at an even lower one, and so on. Importantly, if the plan failed to work at a high level, no energy and time need be spent on working out complicated details at a lower one. Interestingly, when used to program a robot (to open a door, pass through one room to another, push together two boxes, then proceed through another two doors to a final room) this 'high-level' technique took just under 5½ minutes of computer time compared with over 30 minutes taken when the 'high-level' technique was not used. Details will be given in Chapter 12 (STRIPS and ABSTRIPS).

Once bitten, twice shy

What happens if, despite breaking plans into subplans, giving preconditions criticality levels, etc. you still come up against an insurmountable problem when searching for a solution and simply have to abandon that particular line of searching and backtrack?

> THINK!
> - You have decided to spend a week with a friend in Manchester. Although a national rail strike has been threatening for some time, you decide to go by rail. You get a taxi to the station—only to find that the strike is on.
> - You are determined to get to Manchester. What do you do?

This question is far more interesting in showing what you would *not* do than in showing what you *would* do! Obviously you would have no alternative but to arrange for another form of transport.

However, unless told otherwise, a computer may try to solve your problem in a rather less intelligent way. In going back to the point where it ran into problems, and trying out all other possible options to see if they are any more successful, it may, for example, suggest that you try to get to the station by bus instead; failing which, you should walk. However, even having got to the station, if you still cannot board a train it may suggest that you get a taxi to another station within reach. When this plan fails, the computer again suggests travelling to that station by bus, and so on. Only having exhausted all possibilities will the computer 'backtrack' to other choices made even earlier, and suggest travelling to your friend's home by some other means. Hardly intelligent behaviour!

The knowledge *you* used in making such short work of realizing the need to arrange alternative transport related to *consequences*. The consequence of the strike being a national one was that no trains would be running from any station, and whether you took a taxi, a bus, or walked to the station could not affect this situation in any way. Rather than 'backtracking' from the point where you actually experienced problems to the previous point and trying out all possible alternatives from there, you had to rule out *all* the options which depended on the basic choice of travelling by rail. In other words, rather than trying different stations, and different ways of getting to them you take the very much more effective step of 'backtracking' to the very root of the problem, and ruling out *all* explorations relating to trains.

The different approaches are best seen using a diagram such as that in Figure 17. This shows the possibilities open to you. Starting at the top of the figure, you could decide to travel by rail or road. Having decided to travel by rail, you have the possibility of going from one of two stations within reach, and so on.

If you take no notice of the fact that things higher up in the figure have consequences for those beneath them (the *whole* of the figure connected to the 'travel by rail' decision cannot be used if there is a national strike) you may end up looking rather foolish. Like the computer, you may get to point 5 on the figure and simply backtrack to point 6 and try again. On failing to get a ticket *again* at point 7, you may backtrack again to point 8, trying again to buy a ticket at point 9!

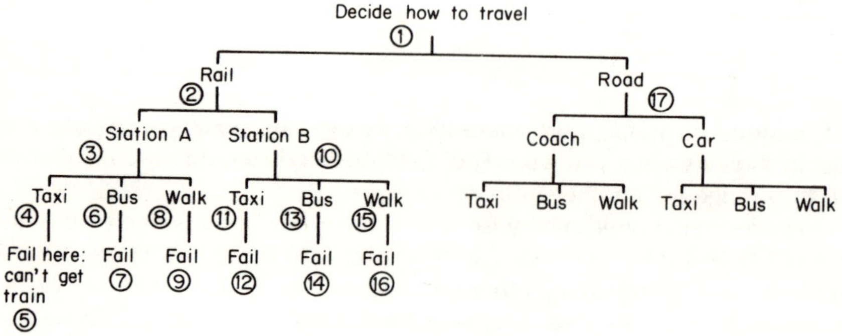

Figure 17. Travel options

If you *do* take notice of what actions and decisions depend on what other actions and decisions you would, on failing to get a ticket at point 5, immediately go to point 17—realizing that a strike 'knocked out' all points from 2 to 16 inclusive, and that exploring any of them would be a waste of time and energy.

Figure 17 looks rather like an upside-down tree—and indeed the jargon-word for cutting down the number of 'branches' that you have to go through is often referred to as 'pruning'!

Human beings can easily remember the *reasons* they have for doing and thinking things. It was not too hard for you to remember that the reason for taking a taxi to the station and trying to buy a ticket was to travel by rail to your friend's house. When the whole reason was removed (you could not travel by rail at all) all the 'branches' connected to that reason were 'pruned'.

In a similar way, we can program the computer to keep track of the *reasons* it has for pursuing various stages of problem-solving. If it records its reasons, it can use this record to backtrack in a more intelligent way. Such a record is termed a 'dependency record', and using such records is called 'dependency-directed backtracking'.

Basically, what is stored in the computer's memory is a series of beliefs, along with the evidence, and inference rules used to process the evidence to produce the beliefs. For example:

Belief:	No trains will run in the country
Evidence:	There is a strike The strike is a national one
Rules:	If an industry is on strike, that industry's services will not be available If a strike is national, it affects the whole country

Using such a dependency record allows more intelligent decisions to be made. It allows more flexible thinking.

THINK!
- Imagine that you had wrongly *assumed* that the strike was national. A porter calls after you, as you leave the first station, that the strike is purely local to this station.
- Looking at Figure 17, backtrack from point 5 to consider what you would do next.

You would be most likely to backtrack to point 10 and try the other station. What you were able to do was *reject* the false assumption that the strike was a national one but *retain* the correct knowledge that there was indeed a strike. Whilst retaining the valid evidence and rules, you could substitute other more relevant ones for the ones that do not now apply (the new ones are in italics):

Belief:	*No trains will run from station A*
Evidence:	There is a strike *The strike is a local one*
Rules:	If an industry is on strike, that industry's services will not be available *If a strike is local, only local services will not be available*

In this way the computer can record not only all the inferences (new knowledge, beliefs, and decisions) it makes but also all the evidence and inference rules used to make them. Often, the same piece of evidence or the same inference rule may support a variety of different beliefs or decisions, depending on the *other* beliefs and rules with which it is combined at the time. By simply jettisoning only those pieces of evidence or rules which are not valid or do not apply, the computer avoids having to go back and start reasoning 'from scratch' each time it encounters a problem. New combinations can be formed relatively easily and flexibly using such a 'belief-revision' system.

Although an assumption that the strike was national would in this case have been incorrect, assumptions can be very useful. They can drastically reduce the amount and complexity of thinking necessary in solving many problems and making many types of decision.

Imagine, for example, that you decide to go shopping. It is a fair assumption that if it is a weekday, and is mid-afternoon, most of the shops you need will be open. It is more intelligent thinking than telephoning each shop beforehand to check that it is open! Usually, intelligent assumptions need some qualification. For example:

> Assume that the shops will be open *provided that there is no evidence to the contrary*

In other words, the assumption may not apply if it is a bank holiday, if you heard rumours of a strike, if the shop you need is a small specialist shop which you have found closed in the past when you expected it to be open, etc.

Much of our lives is based on making such assumptions and knowing when we can and cannot apply them. Computers can also be programmed with assumptions to allow them to avoid unnecessary processing. Assumptions can be expressed as rules, like any other rules:

> If there is no evidence to the contrary, then assume that the shops will be open
>
> If there is no evidence to the contrary, then assume that the buses are running

In this way, assumptions may be made provisional. That is, they are subject to revision in the light of mounting evidence to the contrary. If the bus you are waiting for does not turn up then it is reasonable to assume for the moment that it has broken down. If the next bus does not turn up either then you would begin to doubt your assumption that the problem related only to the first bus. If the third bus failed to appear you would probably reject the assumption and take steps to find out what was going on.

Many intelligent computer systems weigh the balance of evidence supporting a variety of different conclusions and indicate which are the most likely, making use of *probabilities* (which we will look at in the next chapter).

Opportunistic, multidirectional planning

So far, we have looked at very clearly structured ways of solving problems and making plans. 'Backward-chaining' involves seeking evidence to support a hypothesis; 'forward-chaining' involves coming to conclusions as a result of looking at evidence. 'Means–ends analysis' breaks down complex tasks into less complex component ones. However, always the order of actions and decisions is fairly predictable.

But consider the sort of 'problem-solving' going on below:

You plan to catch the 9.30 a.m. bus to town, get off at the supermarket and do the shopping, call in at the bank on your way to a dentist's appointment at 10.30. You can get a bus home at about 11.00, but if you get off at the stop before, you can pick up a newspaper from the local newsagent and some fresh fish from the fishmonger a few doors away.

What actually happens is:

You just miss the bus! Since there isn't another for 20 minutes, you decide to nip to the local newsagent 5 minutes' walk away down the hill. You buy a newspaper.

Since you are so near, you think you'll get the fish from the fishmonger as well—but on second thoughts, the next bus is now due, and you don't want to miss it.

You catch the bus and arrive at the supermarket, but now you haven't as much time as you need before your dental appointment. You decide to get the most essential things, and to return after the dentist, if you feel well enough, to collect the rest and to get to the bank.

You get to the dentist on time. It is not as bad as you feared. You return to the supermarket. Whilst there, you meet a friend who persuades you to go for a coffee, at his favourite café 5 minutes' walk away. You decide that the bank is not essential. You can call there tomorrow.

Having had a good chat, you emerge from the café to see a fish shop a few doors away. You decide to get your fish there, and take a bus all the way home. At least you won't have to walk up the hill from the stop before.

A bit contrived, you may think—but remarkably lifelike when you compare the story with the plans made by people in an experiment by researchers in the U.S.A., on which my example is based. They gave people a map of a small town, and told them to plan out a day's activities which were to include ten tasks such as buying various things from various shops.

The researchers were interested in how people really plan and solve these problems. What was clear was that people don't work out such activities in a logical way then stick to their original plan, come what may. They are a lot more flexible, adapting to changing circumstances that they consider best at the time.

People seem guided by both 'top-down' and 'bottom-up' processes. Top-down processes include our overall plans and goals, aims, and objectives. We decide on what we want or need to do and plan a pattern of activities which we then try to make work. Perhaps a better phrase than 'top-down' would be 'coming from within'. Bottom-up processes—perhaps better called 'coming from outside'—include things that happen to us or that we see and hear. Missing a bus, noticing that there is a fish shop where we didn't know there was one, finding ourselves where we didn't originally plan to be at that time—all these would be 'bottom-up' processes. 'Bottom-up' and 'top-down' processes interact in very complex and constantly changing patterns of problem-solving. We use knowledge at various levels—about how much time we have available and the main things we want to achieve; about times of buses, how long it will take us to do certain things; about where we can buy certain things.

Think now of someone in charge of a complex transport network, such as a region of our rail network. He or she will have knowledge and priorities at various levels (for example, that intercity expresses have priority over local trains; that certain local trains are billed as connecting with main-line trains; that certain trains running from (A) to (B) depend on a prior successful run from (B) to (A) and may require the same locomotive or rolling stock; etc.). Now imagine that things begin to go wrong. A couple of main-line trains break down, or some trains are cancelled due to strike action. Line-repair work closes some lines and requires rerouting via others.

What is clear is that each problem cannot be dealt with in isolation. The rail network is an interconnected, and very complex, whole. What happens in one part may affect what happens in another. Locomotives and rolling stock, if delayed, rerouted, or cancelled may end up in the wrong place at the right time, and the right place at the wrong time for other trains depending on them.

This is a more dramatic example of a very similar problem involving complex interactions of 'top-down' plans and 'bottom-up' events at various levels. Although hard to believe, things we do every day but don't usually regard as a problem (such as understanding people talking or making sense of what we see) share many of the features of the two problems described above.

A computer program designed to solve complex problems such as these cannot rely on simple backward- or forward-chaining, or means-ends analysis. These are not geared to coping with constant unpredictable changes in the problem situation from moment to moment. We need a system which can cope with what is often called *opportunistic problem-solving*, involving complex interactions between knowledge at different levels and between top-down and bottom-up processes, as things happen rather than as prearranged. In this way the problem-solving processes constantly react to constantly changing situations, which they themselves are instrumental in changing.

This sort of problem-solving is also *multidirectional*—planning is going on at several different levels of abstraction at the same time. Subproblems at one level are not isolated, solved, then passed up, duly completed, to the next one, and so on. Things are happening at several levels simultaneously. What happens at one level may affect what happens at several others.

A number of systems designed to cope with such complexities make use of what are known as *blackboard* techniques. The central idea behind the use of a 'blackboard' is to set up a team of 'consultants', each of which can read a central noticeboard which gives details of the current situation relating to whatever problem is being worked on. Each consultant looks for particular features of the current situation which it is designed to respond to and act upon. When the situation is such that a particular consultant can act, it may do so. Interestingly, its actions can include updating the noticeboard—deleting, adding, or changing details of the current situation. So the noticeboard can constantly be read by, and changed by, the team of consultants. In this way they can all be made aware of and respond to changes in the current situation, which they themselves are active in changing.

In Chapter 10 we will build a simple blackboard system, also using dependency records. This is to illustrate the central ideas clearly rather than to show the levels of sophistication that can be built into such systems. However, a more complex use of blackboard techniques in relation to the problem of speech-understanding is illustrated by HEARSAY (Chapter 12).

CHAPTER 8

PROBABILITIES

As we found in Chapter 7, it would be useful if we could instruct the computer to follow up a particular line of argument only if the evidence supporting it reaches a certain level of likelihood or probability. In other words, we need to build into the computer's knowledge some estimations of how *certain* various assertions may be and, given the certainty of these assertions, how *likely* particular conclusions based on them may be.

Using uncertain evidence and assessing the probability of conclusions and advice is an important feature of expert systems. Often, the user may be asked how certain he or she is of a particular fact or how true a particular fact is. For example:

Has your car battery been tested within the last 3 months? Answer on a scale +5 to −5 (+5 = definitely; 0 = don't know; −5 = definitely not).

The user is free to answer using any number from +5 to −5, and the system will then calculate, on the basis of this and other questions, how likely a particular fault may be.

Telling PROLOG to act only if evidence reaches a certain level presents no problems. We can specify, for example:

action(get_new_starter):–	Conclude that a new starter is needed if
fault(faulty_starter,Prob),	The probability of the starter being faulty is stored in the variable Prob
Prob > 0.75,	And whatever is in Prob is greater than 0.75

fault(flat_battery,Prob_1),	And the probability of a flat battery is stored in Prob_1
Prob_1 < 0.35	And whatever is in Prob_1 is less than 0.35

In other words,

Get a new starter if there is a more than 0.75 probability of its being faulty *and* a less than 0.35 probability of the battery being flat.

More difficult is telling PROLOG to calculate what the probabilities are.

Techniques commonly used in expert systems to calculate probabilities include 'fuzzy logic' and 'Bayesian updating'. If you don't feel like going into the details of these processes at this stage, skip to the next chapter. It is not necessary to understand them to carry on from Chapter 9.

Imagine that you have established a rule:

If both symptom A and symptom B are definitely (100%) present
then the fault is definitely (100% sure to be) C

The same sort of rule could apply to cars, people, whatever you like. You can think of your own examples of what A, B, and C might be.

However, now imagine that you are pretty sure that symptom A is present, but not at all sure that symptom B is present—we could estimate our degree of certainty as a percentage, say,

90% sure that symptom A is present
30% sure that symptom B is present

THINK!
- How sure would you now be that the fault is C?
- Express your answer as a percentage:

'I would be .... % sure that the fault is C'

Now before we look at a possible solution, try this one. Imagine that the rule we have established has an 'or' in it instead of an 'and':

If either symptom A or symptom B are definitely (100%) present
then the fault is definitely (100% sure to be) C

Imagine that again we have certainty levels as follows:

90% sure that symptom A is present
30% sure that symptom B is present

THINK!	• How sure would you now be that the fault is C? Is your certainty the same as before? • Again, express your answer as a percentage.

The calculation of probabilities is a contentious business. There is no one correct way of doing it—whichever way you choose may be open to criticism. However, one method commonly used in expert-system development is to use what is known as 'fuzzy logic'. This specifies how we can combine numbers linked by '*AND*', '*OR*', and '*NOT*' (sometimes referred to as Boolean operators) and end up with one figure. Fuzzy logic specifies that:

for numbers linked by *AND*, choose the lowest of them as the overall probability

For numbers linked by *OR*, choose the highest number as the overall probability

So in our first example, the overall probability would be 30%. In our second, the figure would be 90%

This makes some sort of sense. If C is only confirmed if *both* A and B are present, then we can only be as certain of C as we are for the *least* certain of A and B. On the other hand, if C is confirmed if *either* A *or* B is present, and both are present, then we can take the highest—whether this is A or B.

These rules can, of course, deal with more complex combinations. For example:

If symptoms A *and* B *and* E *and not* F, are present
then the fault is G

Let's assume that we have certainty values as follows:

	A	70%
	B	30%
	C	25%
	D	15%
	E	45%
Not	F	90%

We can combine them as in Figure 18.

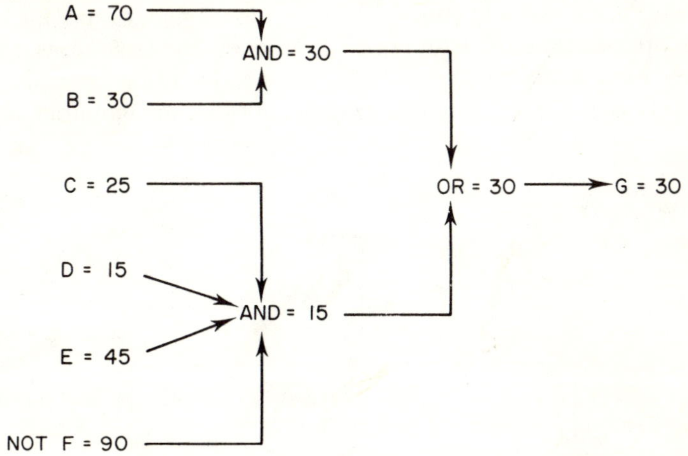

Figure 18. Combined certainties

So far, so good, but we have a problem. Imagine that we are 100% sure of symptoms A and B. According to our calculations so far, this would mean that we must therefore also be 100% sure that the fault is G. This may be all right for some types of rules. For example:

If you have a 2
and add another 2 *then* you have 4

Much knowledge is not, however, so clear-cut as this. In many areas of knowledge—perhaps most—we need to interpret rules like

If A and B *then* C

as saying 'even if we are 100% sure of both A and B, we can still only be 90% sure of C, because of other unknown factors.' We may even need to say 'on the other hand, even if we are 100% sure that A and B are both completely *false*, we certainly mustn't conclude that C is 100%—or 90% for that matter—sure of being false.' It may be that if A and B are completely false, it may tell us that C is likely to be 5% false; or it may even tell us nothing at all about C!

We may also want to take account of cumulating evidence. If a number of factors all point to the same conclusion, then it would seem sensible to give a greater certainty to that conclusion than if it were supported by just one. The greater the supporting evidence, the more certain the conclusion. However, this process is not without its difficulties. Look, for example, at Figure 19. In this figure the AND and OR are fairly clear-cut. The advice is to call the garage if *either* the battery is flat *or* you are out of petrol and have no spare petrol can which you could carry to a garage and fill. The two conditions are separate. The opposite is true of the conditions related by AND. They must *both* be present before the conclusion is valid. In other words, if only one applies then the conclusion (call garage) does not apply.

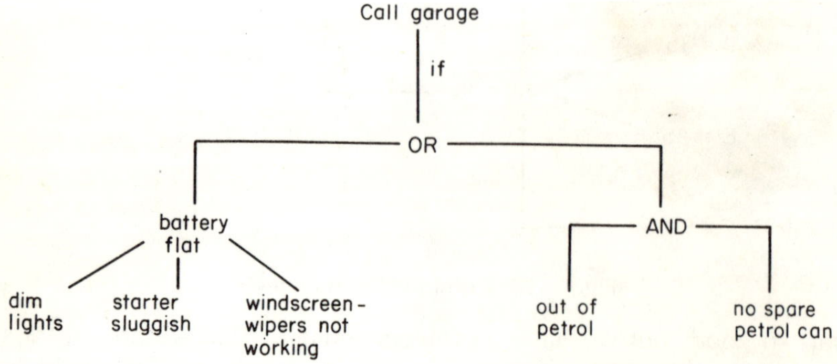

Figure 19. Culminating evidence

The three conditions indicating a flat battery, however, are different. They have a bit of OR about them, and a bit of AND as well! In other words, a flat battery may be indicated by dim lights *or* a sluggish starter *or* windscreen wipers that don't work. On the other hand, the likelihood of the battery being flat is increased the more of these indications are present. The battery is *more* likely to be flat if the lights are dim *and* the starter is sluggish *and* the windscreen-wipers are not working than if only one or two of these symptoms were present. They represent cumulative evidence of a flat battery in a way that the other AND and OR conditions are not cumulative. Also, what if the evidence of a flat battery conflicts—the lights may not be so dim yet the starter may be sluggish (needing more electricity to power it)? Boolean logic stating 'maximum for OR' and 'minimum for AND' do not help here. Which would we choose?

In expert systems a widely used way of getting round these difficulties is to utilize 'Bayesian updating', based on Bayes' Theorem, which relates to calculating probabilities. To use Bayes' Theorem we first need to know the prior odds of whatever hypothesis we are dealing with. Imagine that our hypothesis is that we have a flat battery. The prior odds of this hypothesis relate to the degree to which a flat battery may be expected to occur in the normal course of events. Imagine that, of 100 cars coming into a garage for repair, 25 have flat batteries and 75 do not. The prior odds of having a flat battery would therefore be 25/75. If we do not have such statistics to hand (as is often the case when designing expert systems) we can use our experience and judgement to come up with a figure.

Next, we require some figures to reflect the sort of reasoning we talked about earlier, which says 'even if we are 100% sure of dim lights, this on its own can only make us 55% sure that the battery is flat', etc. We can do this by attaching to each factor two numbers. One specifies the maximum effect on the certainty of the conclusion to which it relates. (Such a figure can be supplied on the basis of experience on the part of the persons building the system.) A complementary number specifies what the maximum effect would be on the conclusion if the factor were completely *false*. For example, we may think that if we definitely know (100%) that the lights are dim, this fact will make the likelihood of a flat battery, say, 25% more certain. If we know definitely (100%) that the lights are *not* dim, this fact will make the likelihood of a flat battery 95% less certain. The first figure is referred to as the logical sufficiency factor (LS), the second as the logical necessity factor (LN).

Assuming for the moment that we are 100% sure of our evidence, we are now in the position to apply Bayes' Theorem to calculate the overall likelihood of our conclusion being true, given this evidence. The formula states that:

$$\begin{array}{c}\text{The odds in favour}\\ \text{of our hypothesis}\end{array} = \text{The prior odds} \times \begin{array}{c}\text{The LS (if the evidence is positive)}\\ \textit{or}\text{ the LN (if it is negative)}\end{array}$$

This formula can deal with cumulating pieces of evidence. The likelihood of the battery being flat increases as more and more evidence from separate sources mounts up. Bayesian updating handles this by simply adding the LS or LN of any additional source of evidence into the formula:

$$\begin{array}{c}\text{Odds in favour}\\ \text{of hypothesis}\end{array} = \begin{array}{c}\text{Prior}\\ \text{odds}\end{array} \times \begin{array}{c}\textit{First}\\ \text{LS or LN}\end{array} \times \begin{array}{c}\textit{Second}\\ \text{LS or LN}\end{array} \times \begin{array}{c}\textit{Third}\\ \text{LS or LN}\end{array} \cdots$$

and so on for more sources of evidence.

The formula so far gives us odds in favour of our hypothesis. We can easily convert odds into probabilities by using the formula:

$$\text{Probability of our hypothesis} = \frac{\text{Odds in favour of it}}{\text{1 plus the odds in favour of it}}$$

Finally—but very importantly—we need to take account of varying levels of certainty in the evidence itself. In our judgement the lights may be 'very' dim or the starter may be 'a bit' sluggish. We would need to convert these judgements into appropriate figures such as 85% and 25%. Some current systems ask the user for a judgement in terms of 'Very', 'A little', 'Don't know', etc., and convert these replies into figures automatically. Conversely, some systems ask the user to give a figure from, say, +5 to −5, indicating how certain they are or how much a factor is present.

We can add such considerations into our formula by calculating a new LS or LN using the formula:

$$\text{New LS or LN} = \text{Original LS or LN} \times \frac{\text{Certainty of}}{\text{the evidence}} + \frac{\text{1 minus the certainty}}{\text{of the evidence}}$$

An example is shown in Figure 20.

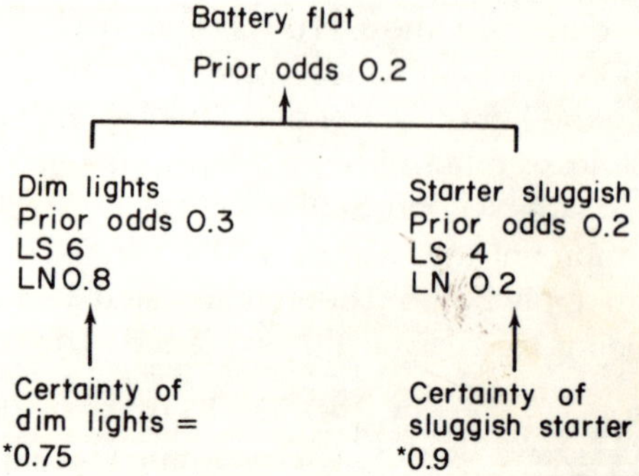

*(These figures would come from the user's replies to the system's questions)

Odds =

Prior odds		New LS for dim lights		New LS for sluggish starter		
0.2	×	(6 × 0.75 + 0.25)	×	(4 × 0.9 + 0.1)	=	3.515

$$\text{Probability} = \frac{3.515}{4.515} = 0.778$$

Figure 20. Calculating probabilities

Bayesian equations are commonly used in expert systems. An example of an expert-system 'shell' (an expert system designed to handle different areas of expertise) which uses them (SAVOIR) is described in Chapter 12.

Although not its best feature, PROLOG can, of course, do arithmetic. However, different PROLOG systems have different levels of mathematical sophistication. I will first give a simple example to show how PROLOG copes with calculating probabilities. In Appendix 2 I will give the details of a small program which, unlike the example, will handle more than one piece of evidence, will not be limited to the 'flat-battery' hypothesis, and will automatically choose whether to apply the LS or the LN depending on the user's answer. However, this program uses decimals and negative numbers. If you are programming with PROLOG, check that your version allows these.

Our example will relate to just one piece of evidence—sluggish starter—relating to one hypothesis—that the battery is flat. It will, for the sake of clarity, assume that the user will reply that to some degree the starter *is* sluggish—in other words, at this stage we will not deal with the problem of having to decide whether to use the LS or the LN. The LS will be used. Now here are some PROLOG rules which will do the calculation for us:

ls(starter_sluggish,battery_flat,4).	The LS of the evidence (sluggish starter) supporting the hypothesis (battery flat) is 4
prior_odds(battery_flat,0.2).	The prior odds of the battery being flat is 0.2
probability(battery_flat, Probability):–	The probability of a flat battery can be determined by finding out
odds_for(battery_flat,Odds_for),	The odds in favour of the battery being flat (which will be worked out by the 'odds_for' rule below, and stored in the variable Odds_for)
X is 1 + Odds_for,	Working out 1 + whatever is in the Odds_for variable, and storing the result in the variable *X*
Probability is Odds_for/X.	Dividing whatever is in Odds_for by whatever is in X, then storing the result in the variable *Probability*. Note that the question which starts the whole program is:

	?– probability(battery_flat, Probability).
	At this stage, PROLOG has worked out what it is, and will reply
	Probability =
	followed by whatever figure is stored in *Probability*
odds_for(battery_flat, Odds_for):–	The odds in favour of a flat battery can be determined by finding out
prior_odds(battery_flat, Prior_odds),	The prior odds of the battery being flat (the 'prior_odds' fact at the beginning of this program tells PROLOG that 0.2 should be stored in the variable Prior_odds)
new_ls(starter_sluggish, battery_flat, New_ls),	The new LS of the sluggish starter as evidence of a flat battery (worked out by the 'new_ls' rule below, and stored in the variable New_ls)
Odds_for is Prior_odds * New_ls.	Then multiplying whatever is in Prior_odds with whatever is in New_ls, and storing the result in Odds_for
new_ls(starter_sluggish, battery_flat,New_ls):–	The new LS of the sluggish starter as evidence of a flat battery can be determined by finding out
ls(starter_sluggish,battery_flat, Ls),	The original LS, which once found will be stored in the variable *Ls*
user_answer(sluggish_starter, User_answer),	The user's estimate of how certain is the fact that the starter is sluggish (stored in User_answer by the 'user_answer' rule below)

New_ls is Ls * User_answer + (1 − User_answer).	Then multiplying whatever is in *Ls* with the result of taking whatever is in User_answer from 1, and storing the total in the variable New_ls
user_answer(starter_sluggish, User_answer):–	To find the user's estimate of how certain the fact of a sluggish starter is
write('How certain are you that the starter is sluggish—answer with a number from 10 (definite) to 0 (don't know)'), nl,	Write on the screen: How certain are you that the starter is sluggish—answer with a number from 10 (definite) to 0 (don't know)
read(Reply),	Store whatever the user replies in the variable *Reply*
User_answer is Reply/10.	Divide whatever is in *Reply* by 10 (in order to convert it into a value from 0 to 1), and store the result in User_answer

We can now ask PROLOG:

?– probability(battery_flat,Probability).

and PROLOG will reply:

How certain are you that the starter is sluggish—answer with a number from 10 (definite) to 0 (don't know)

If we type in:

9 .

PROLOG will reply:

Probability = 0.74

You will find a more detailed program able to calculate probabilities not restricted to a sluggish starter and a flat battery in Appendix 2.

CHAPTER 9

FRAMES OF REFERENCE AND EXPECTATIONS

Dealing with probabilities—deciding on the most likely interpretation of uncertain, ambiguous, or conflicting information—is surprisingly central to human thinking processes. Even at the level of basic perception—seeing and hearing, smelling, tasting, and feeling—we are constantly interpreting information in the light of what we already know. Have a look at the picture shown in Figure 21. What does it show?

You may interpret the pictures as looking up a tall building with birds flying around above it. Alternatively, you could be looking down at people in the street. If you try, you can 'click' from one interpretation to another. The interesting point is, however, that it's difficult to look at the picture without imposing one or other interpretation on it—that is, you do not see it as both looking up and looking down *at the same time*.

The same process may be provoked by giving the same figure two differing contexts, designed to lead you into using one or another part of your existing knowledge to produce different interpretations, as in Figure 22. Human beings find no difficulty in interpreting the figures as '15' or 'IS', as appropriate.

When we can't interpret something as we expect to be able to—that is, in terms of our existing knowledge—we may develop uneasy feelings. Look at Figure 23, for example. You probably looked at the picture, and tried to make some sense of it, wondering just where it went wrong. However, the figure is made up of simple lines drawn on paper. How can lines be 'wrong'? The point is, again, that we try to 'make sense' of them. We try to impose our own interpretation on them to make them conform with our expectations of what ought to be there. They can only be 'wrong' in this interpretative sense. Interestingly, it is very difficult to 'switch off' our interpretative powers. Try,

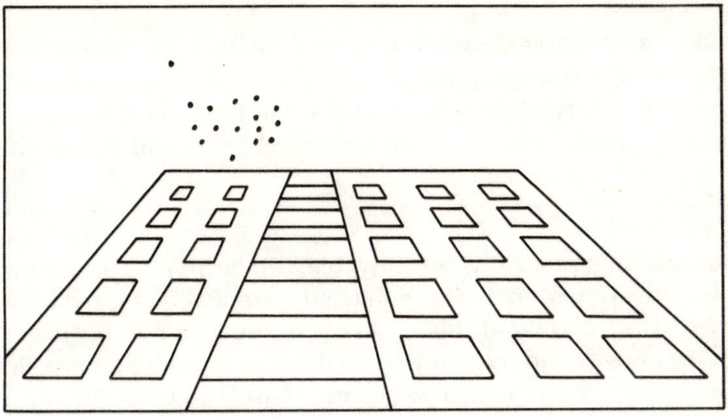

Figure 21. An ambiguous picture

14 IS 16

HE IS A MAN

Figure 22. Different contexts

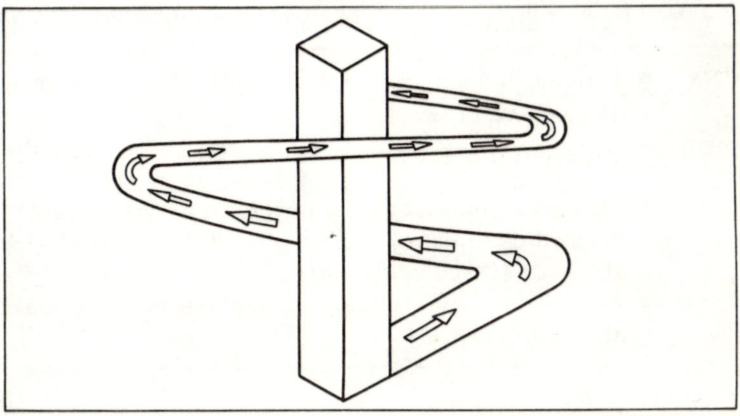

Figure 23. An 'impossible' picture

for example, to look at the figure for what it is—a series of lines on paper—without feeling that anything is wrong.

Making likely interpretations is also a key component of complex intellectual skills—for example, understanding language.

THINK!
- What emotion do you think Mary is experiencing in the following passage?
- 'Mary stared into his eyes. He threw her some chocolate. She hated anything sweet but she ate it without noticing.'

At least two interpretations are possible. Mary may be fascinated by the man—alternatively, she may be frightened, perhaps being held somewhere against her will. If the word 'his' in the first sentence had been replaced by 'John's' most of us would probably reject the 'fear' interpretation as less likely. The point is that in order to understand—make sense of—the passage, we must use our existing knowledge to make likely assumptions about what is really going on.

Psychological research indicates that in perception, language-comprehension, and learning in a broader sense, human beings blend *both* information coming into the senses (words on a page, sounds, pictures) *and* existing knowledge stored in memory in a constant process of generating and testing likely hypotheses. The 'melting pot' where this process takes place is thought to be our *short-term* or *working memory*. So far, when we have been talking about memory we have been concerned with *long-term* or *permanent memory*. What's the difference between short-term and long-term memory?

THINK!
- Look at this number for 4 seconds—then cover it up so that you can't see it:

  1 0 6 6 9 9 9 4 8 9 1

- Make sure the number is covered up. Now count backwards in 4s from 200—e.g. 200, 196, 192, 188, etc.—for 10 seconds or so.
- Now write down the original number that you looked at!

The counting backwards was to occupy your mind and prevent you from repeating the number to yourself over and over again (as we do when trying to remember a telephone number)—a process called 'rehearsal'. In this way, you were relying on the capacity of your short-term memory rather than constantly 'recharging' it through rehearsal.

Now go back and repeat the exercise—but with one important difference. The difference is that I will give you a key to the number. It is in fact made up of three components:

(1) The date of a famous historical battle;
(2) A well-known telephone number;
(3) Another famous date—written backwards.

Armed with this knowledge, you should be much better able to remember the number.

For various reasons, this was a very unscientific experiment, but it does serve to show differences between short-term and long-term memory. Psychological research suggests that we can hold information in our short-term memory for only about 20 seconds, without rehearsal. The amount of information we can hold is also limited, although the precise amount is not clear. However, the numbers 10669994891, if remembered as individual numbers, rather than related groups, should have thoroughly overloaded this capacity! The amount of information and the length of time it can be held in *long-term* memory, on the other hand, is vast. Dates and numbers present no storage problems if stored meaningfully in long-term memory, and it was by drawing on your long-term memory—in terms of well-known dates and numbers—that you could better remember the number the second time you tried.

The point of this little test was to illustrate the importance of the *interaction* between existing knowledge and new incoming information—the interpretation of new information in terms of what is already known. This may come as something of a surprise. It would be easy to imagine that when, say, reading a sentence we simply read the individual words, recognize their meaning, move on to the next words, adding their meaning, until we have built up the meaning of the sentence as a whole. Such an approach may be thought of as 'bottom-up'—building up meaning as we might build a wall, brick by brick, one on top of another. Research suggests, however, that effective reading involves a good deal of 'top-down' processing—guessing what the meaning of words, sentences, and passages might be expected to be, and using the actual words being read to confirm or reject these ongoing guesses. In this way, rather than thoroughly reading each word the effective reader is selecting parts of the text in the light of his or her ongoing expectations about the meaning of what is being read. Both 'bottom-up' and 'top-down' approaches are necessary. The point is that they proceed in close interaction, the one depending on the other.

Many words have a variety of different meanings, as you can see from browsing through any dictionary. If we had to try out each meaning of a word, in combination with each meaning of another word, and the same for the next word, before we hit upon the correct meanings of each of them when combined together, the number of possible permutations would often be enormous. Reading in this 'bottom-up' way would be laborious, if not impossible. What allows us to reduce the task of understanding to manageable proportions is our ability to make enlightened guesses as to likely meanings based on our expectations. In this way, we can read long passages, whole books, full of potentially ambiguous and conflicting meanings, yet make sense of it all and achieve a quick and coherent understanding.

What, however, is the nature of these 'expectations' forming the basis of 'top-down' processing?

Psychological research has explained such expectations in terms of (relatively) self-contained patterns of ideas stored in memory, rooted in past experience. These may be called 'frames', 'scripts', 'schemas', or 'plans'. There are distinctions between these various terms, but essentially what they have in common is the fact that they are fairly self-contained 'modules' of knowledge (nevertheless able to interact with one another) containing stereotypes of various aspects of the world. For example, a module relating to 'travelling by train' will have stored in it the knowledge that normally a traveller will go to a station, often by transport other than a train. He or she will have bought, or will need to buy, a ticket, without which he or she will not be allowed to travel. Trains leave at predetermined times—rather than by personal arrangement—and travellers must find the correct platform. Tickets must be shown at the barrier, etc.

This module will interact with a 'travelling by taxi' module where necessary. Such a module would contain knowledge about the need to telephone in advance and to have a suitable tip ready, etc.

Such pre-stored patterns of knowledge give us certain *expectations* about likely events and situations. These allow us to form appropriate mental contexts—and shut out other inappropriate ones—which help us to reduce ambiguity, and to make sense of what we are seeing, hearing, or reading. In this way, we can quickly understand the following passage:

The train left at 8 a.m. It was 7.45 and John was sweating. He heard a horn outside. He dashed downstairs and turned the key in the lock. The taxi took ten minutes to get there. John only had a £10 note. 'This is a great start!' he said to himself.

THINK! • Give the most likely interpretation of each of the following extracts from the passage:

(1) 'It was 7.45 and John was sweating.'

Was this 7.45 in the morning or evening?
Was John sweating because it was a hot morning or because he was hurrying?

(2) 'He heard a horn outside'.

Was this horn a musical instrument or a car horn?

(3) 'He dashed downstairs and turned the key in the lock.'

Was this the lock of a safe or a door?
If a door, was it a car door or a house door?

(4) 'The taxi took ten minutes to get there.'
Where?

(5) ''This is a great start!' he said to himself.'
Did John mean this literally?

Simple questions—but how did you answer them? What assumptions did you make in order to settle the potential ambiguities? If we can make them explicit, then we are well on the way to being able to equip computers with the power to make likely assumptions based on expectations.

Frames and *scripts* have been used to build assumptions and expectations into computer systems. Frames can represent information about the normally expected characteristics of particular things. For example, a room will normally have a window, four walls, and a door. We know this and we don't find it difficult to understand the following sentences: 'John came into the room. He could see that it was raining more heavily now.' We know that it wasn't raining *inside* the room. We know that John looked out of the window. For a computer to make sense of this it would need to know what we know about rooms. However, there may be cases where a room is still a room but not a typical one. It may not have a window, or it may have five walls, for example. A way round this problem is to make a frame with default values. In other words, we tell the computer that a room has a window, four walls, and a door—*unless we tell it to the contrary*.

Scripts are similar units of information given to the computer so that it can know more than it is explicitly told on a given occasion. These refer to likely sequences of events. If we go to a restaurant, for example, it is normal to enter, sit down, order food, eat it, pay for it, and leave the restaurant. Alternatives can be built in—for example, if it's a self-service restaurant we need to take a tray and queue up. We can also build in a default—unless told otherwise, do or assume such and such a thing. However, the main idea is to have a 'ready-made' sequence of likely events which the computer can call up when appropriate.

Frames and scripts can be 'triggered' by incoming information. For example, from the statement 'John entered the restaurant ...' the words *entered* and *restaurant* could trigger into action a *restaurant* frame. Frames are rather like containers with ready-made slots—imagine a box designed to contain a number of bottles. The bottles themselves may contain very different substances—the same type of box could contain gin or disinfectant bottles. A box may be full or only partially full. Let's take a look at a simple frame and a related script, as shown in Figure 24.

RESTAURANT frame

Kind-of: (Business establishment)
Type: (   )
  range: serve-yourself, waiter-service

  to-find-out: *if* stack-of-trays *then* serve-yourself
  if waiters-at-tables *then* waiter-service
  otherwise serve-yourself

Location: (   )
  To find out: use 'look at menu' procedure → attached procedures

Address: ( )
  To find out: use 'look at menu' procedure → attached procedures

Event-sequence: (   )
  default: Go-to-restaurant Script

GO-TO-RESTAURANT script

Props: restaurant, money, menu, food

People: customer, proprietor

Event-sequence:
  first: Enter-restaurant Script
  then: *if* waiter-service *then* Take-a-seat Script
  otherwise Queue-for-service Script
  then: Order-food Script
  then: Pay Script
  then: Leave-restaurant Script

Figure 24. A frame and related script. (Based on examples given in Frames and scripts, by James Bennett. In *The Handbook of Artificial Intelligence*, vol. 1, Avron Barr & Edward A. Feigenbaum (eds), Los Altos: William Kaufmann, Inc., 1981, p. 217–222.

In this figure the slots are shown by brackets. The 'kind-of' slot has already been filled in—we know that a restaurant is a kind of *business establishment*. The other slots are not yet filled—shown by empty brackets (   ). However, attached to these slots can be procedures telling the computer how to go about filling them. To fill the 'type' slot the computer is told to check whether there is a stack of trays (i.e. whether a statement to this effect is contained in its memory). If there *is*, then the 'type' slot can be filled with *serve-yourself*. On the other hand, if there were waiters at any of the tables then the slot would be filled with *waiter-service* instead. To fill the 'location' slot, the computer is told to use the 'look at menu' procedure—this is a linked procedure (not shown in detail here) which the computer can use to find the location of the restaurant and fill this slot.

The 'event-sequence' slot has a 'default' procedure. This means that unless there are any instructions to the contrary the computer should fill this slot with the events shown in the related GO-TO-RESTAURANT script (shown below the RESTAURANT frame). This script lists the normally expected sequence of events which, unless told otherwise, the computer should assume will take place. The 'props' and 'people' slots at the beginning of the script refer to other frames. In other words, there will be separate frames for RESTAURANT (the one we have been looking at), MONEY, MENU, FOOD, CUSTOMER, and PROPRIETOR.

In this way a computer program may know, and be able to answer questions on, much more than it appears to have been told. For example, a program called SAM (Script Applier Mechanism) developed by Roger Schank and Robert Abelson at Yale University is able to 'understand' stories—to interpret, summarize, and expand on statements—by matching the sentences which make up the stories with the events in one or more scripts. For example, from the following sentences:

John went to a restaurant. He sat down. He got mad. He left.'

SAM produced the following summary (quoted in Barr and Feigenbaum, 1981, Vol. 1, p. 312):

> John was hungry. He decided to go to a restaurant. He went to one. He sat down in a chair. A waiter did not go to the table. John became upset. He decided he was going to leave the restaurant. He left it.

It based this interpretation on its knowledge that people are likely to get upset if what they expect to happen does not. A *go-to restaurant* script (more detailed than the one we looked at) told SAM that the waiter could be expected to come to the table. Since it had no report of this happening, SAM inferred that this was why John became upset and left. (Incidentally, SAM can produce summaries in English, Russian, Spanish, Chinese and Dutch.)

THINK!

- Imagine that you have only a short amount of time to summarize—in only two lines at the most—the report of a football match shown below. Your summary will appear in tonight's local paper.
- Working as fast as you can, produce a two-line summary and write it down.

Barnsley certainly pleased their manager Allan Clarke when they defeated Doncaster at a wild and windy Oakwell on Tuesday night. Previously they had had a roasting from Clarke for losing to Shrewsbury the week before. The first goal came from Walsh after 25 minutes. He hit a screamer that went in off the foot of the post. Paul Futcher scored the second and third goals—both brilliant individual efforts—to make it a 3–0 victory by Barnsley. It would have been more but for the splendid efforts of goalkeeper Steve Perks.

Goals: Walsh (25 minutes), Futcher (50, 55 minutes).
Attendance: 5662.

What you have had to do is decide which are the really essential points to include and which are less important and able to be left out. Most people would agree on the essential points. You will probably have them in your own summary. They include:

Which teams were playing; where they were playing; which team won; what the score was; and who scored.

THINK!

- One way of interpreting what you have just done is to say that you were using an internal *frame*—a ready-made knowledge of the important features to look for with *slots* waiting to be filled. Filling the slots guided the way you read the passage. Probably the first things you looked for were the names of the teams playing, for example.
- Now write out a *frame*—in the same sort of style as the 'restaurant' frame shown previously in Figure 24—which a computer could use to produce summaries of most football matches from longer reports, just as you did, but don't bother with any 'default' or 'to-find-out' instructions yet.

The frame I came up with is shown in Figure 25. I dare say that yours is similar. The main thing, though, is that it does the job. It will need to include the key elements that the summary will need to mention. These elements should be named in a general sense—*teams, score, scorers*, etc.—and should have empty spaces waiting to be filled.

As we saw earlier, we can attach to particular slots instructions to the computer on how to fill them. In other words, we can give the computer procedures to use which will enable it to find the information it needs to fill the slot. Think back to how *you* found the relevant information.

```
FOOTBALL-MATCH frame

  Place: (        )
  Teams: (        )
  Winners: (        )
  Score: (        )
  Scorers: (        )
```

Figure 25. A frame for summarizing football reports

It sounds a simple task. You knew who were playing, who won, and by how many goals because the report told you. However, in understanding the words so effectively you made use of a tremendous amount of knowledge about language and about sport. Otherwise you *could* think that Clarke was a member of—even the name of—another team ('Previously they had had a roasting from Clarke . . .'); or that the goal-posts could kick and play football ('hit a screamer from 20 yards that went in off the foot of the post'). Ridiculous! However, avoiding even these howlers is beyond the capabilities of a computer which has not been given the knowledge needed to make sensible interpretations.

The task of providing a computer with wide-ranging knowledge enabling it to read and interpret as we do would be a vast one indeed. One much simpler approach is to give it much more limited—but very specific—guidance on how to find relevant, correct information. Using a frame system this guidance can be attached to the slots requiring filling. In the case of our sample report, procedures similar to those shown in Figure 26 would be sufficient.

These particular 'slot-filling' procedures are not particularly sophisticated in terms of linguistic analysis! In fact, they would probably perform extremely badly if used on other examples of football reports. However, they do serve to illustrate the basic idea. The computer could produce a standard, pre-written summary using the information in the filled-in slots of the frame. For example, it could be told to print the following for any summary involving a football match:

Teams met at *Place* . The match was won *Score* by *Winners* .
The goals were scored by *Scorers* .

The words appearing in italics—*Teams, Place,* etc.—are variables, like the (X) and (Y) we met in previous chapters. In other words, the computer will print in their place whatever is stored in them at the time. This information it will find by referring to the FOOTBALL-MATCH frame slots, where the relevant information has been filled in (in the brackets attached to each slot).

Place: ( )	
to-find-out:	find the first occurring place-name immediately following the word 'at'
Teams: ( )	
to-find-out:	find the first two place-names mentioned in the report
Winners: ( )	
to-find-out:	find a place-name following words like 'win', 'victory' 'beaten', 'defeated by'
Score: ( )	
to-find-out:	find two numbers of less than 20 separated only by punctuation (e.g. a dash) rather than words
Scorers ( )	
to-find-out:	find a personal name nearest to words indicating a particular minute. Otherwise, find names of persons nearest to words like 'goal' or 'score' excluding name occurring closest to the word 'goalkeeper' or 'keeper', or following words like 'saved by'

Figure 26. A football-match frame

The summary produced from our example newspaper report would read:

> Barnsley and Doncaster met at Oakwell. The match was won 3–0 by Barnsley. The goals were scored by Walsh and Futcher.

Without much other knowledge about language and sport the computer would obediently follow these rules—even if they led to a nonsensical report. Much thought and testing would have to go into producing rules which would give consistently effective results for a variety of different reports, worded in different ways.

This example is a very simple one. However, when we examine AM in Chapter 12 we shall see frames being used to produce sophisticated and valuable hypotheses in complex subjects including mathematics, geometry, war-gaming, and the design of integrated circuits.

CHAPTER 10

A SIMPLE BLACKBOARD SYSTEM

We can now build a simple system to illustrate some of the ideas behind frames, opportunistic reasoning using blackboard techniques, and dependency records. The use of frames in a more complex context is illustrated by AM (Chapter 12). Similarly, the more complex use of opportunistic reasoning and blackboard techniques is described in relation to HEARSAY, also in Chapter 12.

In order to illustrate basic ideas clearly the system has been kept very simple. At this level, it appears as an augmented forward-chaining production rule system, and shares some of the features of the systems we have already looked at; indeed it is based on these elements.

However, each set of production rules exists within a structure—an 'expert'. These experts are relatively independent units, represented in simple frames, equipped with knowledge in the form of sets of production rules and procedures. Each expert can read and respond to a 'blackboard'—in this case simply PROLOG's memory. Each looks for certain information on the blackboard. When it finds it, its procedures are activated, it draws conclusions, and, as a result, modifies the blackboard. The new state of the blackboard is then available to be read by any of the experts (including the one that has just acted). These, in turn, may make further modifications, and so on. Problem-solving is opportunistic in that it is largely asynchronous. In other words, the order in which the experts are activated is neither prearranged nor important—each one acts when conditions are right for it to act, when the current state of the problem is appropriate to its particular knowledge.

Blackboards are generally used in complex systems. The experts may constitute 'expert systems' in their own right, equipped with complex and varied knowledge and procedures. Blackboards may also be complex structures divided into specific areas dealing with particular aspects of the problem being considered. Opportunistic reasoning may blend top-down and bottom-up processing, and make use of sophisticated methods for finding the most efficient way forward at any given time.

Blackboard techniques are particularly suited to fast-moving situations requiring flexible reactions to constantly changing circumstances. Let's take a military application. Although it is a firmly tongue-in-cheek example, blackboard techniques are highly appropriate for the development of complex defence systems. However, as previously noted, this example is designed to illustrate a few central ideas in as simple a way as possible.

The system's job is to keep a constant lookout for signs of a missile attack and offer appropriate advice. We will have four experts, each with a specific task. These are:

(1) Confirm whether an attack is taking place, based on reports from two radar stations, either, both, or none of which may be positive (i.e. missiles detected). This consultant will report that an attack is confirmed as definite if both radar 1 and radar 2 are positive, but that an attack is confirmed as possible if only one radar is positive.
(2) Monitor how likely an attack is at any given time, based on certain observations. To keep things simple, our system will take account of only two conditions:
    (a) Whether or not international relations are bad;
    (b) Whether or not certain danger signs have been observed.
(3) Put missiles on alert. Two levels of alert will be possible:
    (a) Red alert;
    (b) Amber alert.
    Missiles will be put on a red alert if there is a high expectation of attack, amber alert if there is a medium expectation of an attack.
(4) Advise appropriate action. Again, to keep things simple, only two courses of action will be advised:
    (a) Get on the hot line;
    (b) Fire missiles.
    Which advice is chosen will depend on the circumstances. If an attack is confirmed as definite, and there is already a high expectation of one, then 'fire missiles' will be recommended. If there is a high expectation of an attack but an attack is only confirmed as possible, then the advice will be to get on the hot line.

These experts will be represented in frames—containers with slots only the contents of which will be different. Our basic frame will consist of the features shown in Figure 27.

The same *type* of information will be contained in each slot for each consultant. However, the details of that information will be different for each

```
expert(job(          ),
    action_conditions(          ),
    procedures(                    ).
```

Figure 27. An 'expert' frame

one. The 'job' slot will be filled with the name of the task of each consultant. The 'action_conditions' slot will contain two main elements: the possible conclusions that the consultant can come to and the evidence which it needs in order to be able to do so. Finally, the 'procedures' slot contains instructions to write messages on the screen, drawing attention to important conclusions or advice—these messages are 'ATTACK STATUS:', 'ALERT', and 'PRIORITY MESSAGE'. The main purpose in including them in this form is to show how procedures (instructions for doing certain things) can be contained in the slots of frames themselves. More complex procedures could, for example, set off a (backward-chaining) search for more information pertinent to particular hypotheses suggested by the state of the blackboard.

Now here are the four experts:

```
expert(job(confirm_attack),
    action_conditions([attack_confirmed(definite),
                          [radar_1(positive),radar_2(positive)],
                       attack_confirmed(possible),
                          [radar_1(positive)],
                       attack_confirmed(possible),
                          [radar_2(positive)]]),
    procedures(write('ATTACK STATUS:  '))).
```

```
expert(job(expect_attack),
    action_conditions([attack_expected(high),
                          [international_relations(bad),danger_signs(observed)],
                       attack_expected(medium),
                          [international_relations(bad)],
                       attack_expected(medium),
                          [danger_signs(observed)]]),
    procedures(write('ATTACK STATUS:  '))).
```

```
expert(job(put_on_alert),
    action_conditions([missiles_on_alert(red),
                          [attack_expected(high)],
                       missiles_on_alert(amber),
                          [attack_expected(medium)]]),
    procedures(write('ALERT'))).
```

```
expert(job(advise_action),
    action_conditions([fire_missiles,
                [attack_confirmed(definite),attack_expected(high)],
            get_on_hot_line,
                [attack_confirmed(possible),attack_expected(high)]]),
    procedures(write(' P R I O R I T Y   M E S S A G E '))).
```

Note the format of the action_condition information. It consists of lists within a list. Take the first example:

```
action_conditions([attack_confirmed(definite),
                    [radar_1(positive),radar_2(positive)],
                  attack_confirmed(possible),
                    [radar_1(positive)],
                  attack_confirmed(possible),
                    [radar_2(positive)]]),
```

There is one 'outer' list, containing everything that is within the very first and the very last square brackets. Within this outer list there are a number of elements:

[A conclusion	attack_confirmed(definite)
A list of evidence	[radar_1(positive),radar_2(positive)]
A second conclusion	attack_confirmed(possible)
A second list of evidence	[radar_1(positive)]
A third conclusion	attack_confirmed(possible)
A third list of evidence	[radar_2(positive)]]

Although the second and third lists each contain only one piece of evidence, they are still lists. The point of having lists is that, if we wish to, we can include any number of pieces of evidence without having to change the basic framework.

Recall that we can 'get at' the head and the tail of a list by referring to them as follows:

```
[Head|Tail]
```

We can easily extend this to stipulate that we want to 'get at' a head which consists of two elements as follows:

```
[A,B|Tail]
```

This simply means that we are referring to a list with a head consisting of two elements, A and B, and a tail consisting of all the rest. If we now cycle through a list recursively, using this new representation, we can break a list into a head (consisting of the first two elements) and a tail, then break the tail into a new head (consisting of the next two elements) and a new tail. We can then break this new tail into a head (consisting of the next two elements) and a tail, and so on until the list is empty.

We will find this useful, since our 'outer list' is made up of pairs of elements. The first of each pair is a *conclusion*, the second is a list of pieces of *evidence*. What we will do is say to PROLOG:

Take the outer list. Take the head of the list which consists of the first two elements (a conclusion and a list of evidence). Check all the evidence, and if it is successfully checked, then draw the conclusion. (In other words, check the second element, then assert the first element.) When you've done this, break the tail into a new head (consisting of the next conclusion and its list of evidence) and do the same. And so on until the outer list is empty.

In this way a consultant can process all the possible conclusions which it is 'authorized' to draw. If all the evidence supporting a particular conclusion is found to be present, then the consultant will assert the conclusion. If all the evidence is not present, then the conclusion will be ignored, and the next one tried, and so on. When a consultant 'draws' a conclusion it 'writes' it on the 'blackboard', which consists of facts stored in PROLOG's memory. For example:

```
fact(radar_1(positive)).
fact(radar_2(positive)).
fact(attack_confirmed(definite)).
```

Facts will be either information typed in by the user or information added by the consultants. Recall that we can add things to memory by telling PROLOG to:

```
assertz(fact(attack_confirmed(definite))).
```

When the system begins to run it will, of course, have no facts to go on. So the first thing we will do is arrange for it to get some. We can do this by asking the user to type in some observation about the current situation—for example, if either of the radars is positive, if international relations are bad, etc.:

ask_for_data:–	To ask for data
write('Type an observation, or "end" to end'), nl,	Write 'Type an observation, or "end" to end',
read(Answer),	Store the user's answer in a variable called *Answer*
not(Answer = end),	Check that the answer is not "end"—if it is, then fail; if it is not, then continue

```
assertz(fact(Answer)).          Add to memory (the 'blackboard')
                                  whatever the user's answer
                                  was, as a fact—e.g.

                                      fact(radar_1(positive)).
```

This rule will ask the user to type in a fact. We could give him or her various options to which the reply would be 'yes' or 'no'. For example,

Is radar 1 positive?
Is radar 2 positive?
etc.

Instead, to keep things simple, we will assume that our user knows that he or she must type in one of a number of options (anything could be typed in—but unless it is one of the options, no conclusions will be drawn from it!). These options are:

```
radar_1(positive).
radar_2(positive).
international_relations(bad).
danger_signs(observed).
```

The user must type the full stop to let PROLOG know that the answer is finished.

Now we need to tell PROLOG what to do with the observation—to see if any conclusions can be drawn on the basis of it. We'll call this procedure 'fire_rules':

```
fire_rules:-                          To fire the rules,

  expert(job(J)),                     Find an expert and call its
                                        job J,

      action_conditions([A,B|C]),       its first conclusion A, its
                                        first list of evidence B, and
                                        all the rest (the tail) C, and

    procedures(P)),                     its procedures P

  checkout([A,B|C],P),                Process A, B, C, and P by using
                                        the 'checkout' rule below—
                                        this rule will check the
                                        evidence and if it is present,
                                        assert the expert's conclusions
                                        and apply its procedures

  fail.                               Having done so—now fail. In
                                        other words, having checked
```

	out one expert, go on and try the next expert, and so on until you run out of experts. At this stage, PROLOG will go on to the next 'fire_rules' rule below
fire_rules.	This rule succeeds without doing anything. In other words, PROLOG has run out of experts to check out, so its task is now successfully completed.
checkout([A,B|C],P):–	To check out *A*, *B*, *C*, and *P*,
check(A,B,P),	First 'check' *A*, *B*, and *P* (in other words, the first conclusion (*A*) and list of evidence (B), and procedures (*P*)
checkout(C,P).	Then do the same to the tail (*C*) and the procedures (*P*). The list is processed recursively in this way, being broken into a new head and tail in each cycle, until the list is empty—at which point the second 'checkout' rule below is tried
checkout([],P).	This rule succeeds without doing anything, when the list is empty. At this point, 'checkout' has succeeded, and control can be handed back to the last line of the first 'fire_rules' rule
check(A,B,P):–	To check *A*, *B* and *P*.
not(fact(A)),	First make sure that *A* (which is a conclusion) has not already been drawn. If it has, it will be present as a fact on the blackboard. In this case, this 'check' rule will fail, and the

	next 'check' rule will be tried. (This second 'check' rule will succeed, but although the original instruction to check(A,B,P) will have succeeded, none of the actions on the next lines of the *first* 'check' rule (the one we are in now) will be performed. What we are saying is 'If this conclusion has already been drawn, don't bother finding out if its evidence is present, and adding the same fact again. Instead, ignore it and try the *next* conclusion.')
check_evidence(B),	Process the list of evidence (*B*) using the 'check_evidence' rule below, then if this is successfully completed,
assertz(fact(A)),	Add to memory (the blackboard) the conclusion (*A*) as a fact, then
call(P), nl,	Apply the procedures (*P*). 'call(P)' means 'set up *P* as a goal'. If *P* were a fact, this would mean checking that the fact existed in memory. If it is an instruction, like 'write', then setting it up as a goal has the effect of obeying the instruction. Thus if *P* contains write('ATTACK STATUS') 'call(*P*)' results in ATTACK STATUS being written on the screen
write(A), nl,	Then write whatever conclusion is in *A*—e.g.

attack_confirmed(possible)

(Both 'call(P)' and 'write(A)' had the same effect of writing on the screen. Both methods were used here just to show different approaches.)

fire_rules.

And apply the 'fire_rules' rule in order to draw any further conclusions on the basis of the newly asserted conclusion *A*. *Note*: this is a recursive use of 'fire_rules'. 'fire_rules' here is being used within 'check', which in turn is being used within 'checkout', which is being used within 'fire_rules'.

The 'outer' use of 'fire_rules' instructs the experts to check the blackboard in order—first expert, second expert, and so on. However, if an expert were to draw a conclusion which an earlier expert could use it could not use it in this 'fire_rules' cycle.

The recursive use of 'fire_rules', however, means that there can be a complete 'chain-reaction' whereby conclusions are drawn, and any conclusions derived from them are drawn, and so on.

When the outer 'fire_rules' results in a conclusion being drawn, this conclusion will set off its own 'fire_rules'. Each conclusion drawn within this 'fire_rules' will set off its own 'fire_rules', and so on. Each level of 'fire_rules' will succeed when all the experts

	have examined and, if necessary, acted upon the blackboard, until finally the outer 'fire_rules' also succeeds.
check(_,_,_).	This 'check' rule will succeed if the first one fails. This success is needed to allow PROLOG to try all the conclusions and evidence in the list, rather than stopping after the first failure
check_evidence([H|T]):–	When told in the third line of the first 'check' rule to check_evidence(B) PROLOG breaks whatever is in *B*, e.g. [radar_1(positive), radar_2(positive)] into a head (*H*) radar_1(positive) and a tail (*T*) radar_2(positive)
fact(H),	Check that whatever is in *H* is present as a fact in memory (i.e. on the blackboard)
check_evidence(T).	Then go on to do the same with everything in the tail of the list (*T*). Although there is only one item in the example list, this rule could cope with any number. This rule will be applied recursively until the list is empty, at which stage the second 'check_evidence' rule below will be tried

check_evidence([]).	When the list is empty, 'check_evidence' has succeeded, and control is returned to the fourth line of the first 'check' rule.

Now that we have all the components of the system all we need are some instructions to apply them in the right order and to start the system off.

We'll call the rule containing these instructions 'lookout'. What 'lookout' will do is go round in a constant loop, asking the user for observations ('ask_for_data'), inviting each expert to examine the blackboard to see whether any of its conclusions can be drawn, and if so, to act accordingly ('fire_rules'), then to cycle through the whole process again and again. It will only stop doing this if the user, when asked to type in an observation or 'end', does in fact type 'end':

lookout:–	To 'lookout'—or in other words, when the user types ?– lookout.
nl,	Go onto a new line, then
ask_for_data, nl,	Apply the 'ask_for_data' rule, then
fire_rules, nl,	Apply the 'fire_rules' rule, then
lookout.	Start again from the beginning. This cycle will only end if 'lookout' fails. 'lookout' will fail if any of its lines fail. The 'ask_for_data' rule in line 3 is designed to fail if the user types 'end'. When this does happen, PROLOG will try the second 'lookout' rule below
lookout:–	'lookout' will have succeeded if
write('Program finished').	PROLOG writes Program finished on the screen.

Now when the user types:

?– lookout.

the system will respond:

Type an observation, or 'end' to end

Assuming the following responses on the part of the user, the consultation would go as follows:

User:	international_relations(bad).
System:	ATTACK STATUS: attack_expected(medium) ALERT missiles_on_alert(amber) Type an observation, or 'end' to end
User:	radar_1(positive).
System:	ATTACK STATUS: attack_confirmed(possible) Type an observation, or 'end' to end
User:	danger_signs(observed).
System:	ATTACK STATUS: attack_expected(high) ALERT missiles_on_alert(red) P R I O R I T Y M E S S A G E get_on_hot_line Type an observation, or 'end' to end
User:	radar_2(positive).
System:	ATTACK STATUS: attack_confirmed(definite) P R I O R I T Y M E S S A G E fire_missiles Type an observation, or 'end' to end
User:	end.
System:	Program finished Yes

The responses of the system, and the advice given, would, of course, vary, depending on the observations about the current situation being typed in by the user. Only a few conclusions and pieces of evidence have been included in this example, but the same framework could be used with many more.

However, one of the main limitations of the system as it stands is that we cannot retract any observations, along with any conclusions that have been drawn which are based on them. We looked at 'dependency records' in the context of 'dependency directed backtracking' in Chapter 7. We could do with such a dependency record in this system.

We can build a simple dependency record by adding an instruction to our existing program. Each time the system draws a conclusion and adds it to the blackboard, we will make it also create a separate record of the conclusion and the evidence it used to draw it. The instruction is simply:

assertz(dependency_record(A,B))	Add to memory whatever is in *A*—e.g. attack_confirmed(definite) and whatever is in *B*—e.g. [radar_1(positive), radar_2(positive)] prefaced by the words dependency_record

This would result in the following record being added to memory:

```
dependency_record(attack_confirmed(definite),[radar_1(positive),
   radar_2(positive)]).
```

A similar record would be added for each conclusion that is drawn.

We will place this instruction at the end of the 'check(A,B,P)' rule, which will now read:

```
check(A,B,P):-
   not(fact(A)),
   check_evidence(B),
   assertz(fact(A)),
   call(P), nl,
   write(A), nl,
   assertz(dependency_record(A,B)).
```

Now that we have a dependency record for all the conclusions that are drawn, listing all the evidence each one was based on, we need a mechanism to exploit it—to allow the user to retract observations and have all conclusions based on it automatically removed from the blackboard. We can add a line to the end of the 'ask_for_data' rule which will call up such a procedure:

```
ask_for_data:–
    write('Type an observation, or "end" to end'), nl,
    read(Answer),
    not(Answer = end),
    assertz(fact(Answer)),
    retract_observation, nl, nl.
```

When PROLOG gets to the last line, it will look for a rule 'retract_observation'. So we must give it one:

retract_observation:–	To retract an observation,
write('Type an observation you wish to retract, or "n" if none), nl,	Write 'Type an observation you wish to retract, or "n" if none'
read(Observation),	Store the user's reply in a variable called *Observation*
(Observation = n;	This rule succeeds if the user's reply is "n" *OR* (the semicolon means OR)
(retract(fact(Observation)),	Delete from memory—the fact that the user wants to retract, and go on to
prune_dependencies(Observation)).	Get rid of those conclusions that depend on whatever is in *Observation* by applying the rule prune_dependencies in relation to *Observation*

Now we need to give PROLOG the 'prune_dependencies' rule:

prune_dependencies(Observation):–	To get rid of the conclusions depending on the user's retracted observation stored in the variable *Observation*
dependency_record(A,B),	Find the first dependency record
belongs_to(Observation,B),	Check whether the user's retracted observation is contained in the list of

	evidence (*B*) supporting the conclusion (*A*)—we will define the 'belongs_to' rule below
process(Observation,A,B),	Send the contents of the variables *Observation, A,* and *B* to be processed by the 'process' rule below (this rule will retract *A*, then retract any second-level conclusion which depends on *A*, then any third-level conclusion which depends on the second-level conclusion, and so on until there are no more levels of conclusion derived originally from *A*
fail.	Fail—in other words, having done all this for *A*, try to find any other conclusions which also depend on the user's retracted observation stored in *Observation*. Only when PROLOG has run out of dependency records will it go on to the next 'prune_dependencies' rule below
prune_dependencies(_).	When PROLOG gets to this rule it will have deleted all conclusions derived from the user's retracted observation. At this stage, 'prune_dependencies' has succeeded
process(Observation,A,B):–	To 'process' the contents of *Observation, A,* and *B*,
retract(fact(A)), nl,	Delete conclusion *A* (which was stored as a fact in PROLOG's memory)

write(A),	Write conclusion *A* on the screen,
write(' : now retracted'),	Followed by the words : now retracted
prune_dependencies(A).	Then apply the whole 'prune_dependencies' rule again, substituting whatever is in *A* for what is in *Observation*. This has the effect of retracting any second-level conclusion which depends on *A*, then any third-level conclusion which depends on the second-level conclusion, then any fourth-level conclusion which depends on the third-level conclusion, and so on.

Finally, we need rules telling PROLOG how to find out whether an observation 'belongs_to' a list of observations. We can do this by including rules which will find out if any item belongs to a list of items—very useful rules generally:

belongs_to(Item,[Item|Tail]).	An item belongs to a list if it is the same as the head of the list
belongs_to(Item,[Head|Tail]):– belongs_to(Item,Tail).	An item belongs to a list if it belongs to the tail of the list

These two rules work recursively—if the item being asked about is not the head of the list then the tail will be broken into a new head and tail, and the first rule will be tried again to see if the item is the head of the new list. If it is not, this tail will be broken into a new head and tail—and so on until either the item being asked about *is* the same as the head or both rules fail. For a 'working' illustration of 'belongs_to', see Appendix 1 Program B.

The user will start the system as before, by typing:

```
?– lookout.
```

If he or she wishes at any stage simply to retract information without adding any, a full stop only may be typed when the system asks:

System: Type an observation, or "end" to end

User:

System: Type an observation you wish to retract, or 'n' for none

The system is, of course, very limited—not only in relation to the knowledge it uses but also the facilities it offers. Its limitations include the fact that it is not possible to specify conditions that should be *absent* before a conclusion should be drawn. For example:

Get on the hot line if an attack is confirmed as definite but there is no evidence of bad international relations.

Nor does the system offer any explanation to the user as to why it has reached a particular conclusion. A slightly more sophisticated version of the system which does provide these facilities in a simple form is included in Appendix 2.

CHAPTER 11

LEARNING

Simple learning

So far, so good. We have seen that the computer is very good at obeying rules, whether to plan a journey, write a football report or keep a lookout for missiles! However, there is more to intelligent behaviour than simply obeying rules, however intelligent the person who makes up the rules may be. The budding tennis- or chess-player will not progress very far by only following his or her coach's instructions to the letter and doing no more! Being taught—however well—is only a springboard, not a substitute, for development. Life is so varied that we cannot rely on only having to cope with situations that we have been explicitly taught how to handle. What we must be able to do is learn for ourselves and build up our own experience.

Learning is a complex business. For the sake of clarity, let's take a simple example familiar to us all—learning about a game! Normally when we learn about a game someone who is already familiar with it will probably explain the ideas and the rules to us. When we are learning things by ourselves, often we have to pick up ideas and rules from our own experience. In other words, we are faced with 'raw' information and have to make sense of it and interpret it for ourselves. At a complex level, if we play chess, poker, or whatever, we might think about our most recent game and wonder why we lost. If we won, we might try to work out what it was that led to success. We might observe players who are better than we are, and again try to work out the 'rules' they seem to be using which might improve our own game.

THINK!
- Imagine that you are going to teach a friend (or a computer) to learn what a *flush* in poker is.
- You mustn't simply tell your pupil what a flush is. Instead you will present a number of *examples* of flushes, and your human or machine companion must learn for him/her/itself simply by examining the examples. A set of examples is shown below.
- Now think. Exactly what instructions would you give to your pupil as to how to go about learning what a flush is—without actually saying what one is? (If you don't play cards and don't know what a flush is then try working it out from the examples. Then think about how you would teach someone else to do what you've just done.)

Here is a set of examples:

(1) A flush	5 HEARTS 3 HEARTS 6 HEARTS 2 HEARTS 7 HEARTS
(2) A flush	5 DIAMONDS 2 DIAMONDS 3 DIAMONDS 4 DIAMONDS 7 DIAMONDS
(3) *Not* a flush	3 HEARTS 5 HEARTS 6 HEARTS 7 DIAMONDS 2 DIAMONDS
(4) A flush	7 CLUBS 3 CLUBS 5 CLUBS 4 CLUBS 2 CLUBS
(5) *Not* a flush	4 DIAMONDS 3 DIAMONDS 6 HEARTS 2 DIAMONDS 7 CLUBS

Here is the advice I would give:

First, find out those features which all the examples of flushes *share*. Second, from this list of features, rule out those which are *also* shared by either of the 'non-examples,' (3) and (5).

The examples of flushes—(1), (2) and (4) all share a number of features. For example, they all contain 2, 3, 5, and 7; they all consist of cards with values which are less than 8; and they all consist of cards of a common suit (all clubs, all diamonds, etc.).

If we take these shared features as our 'short-list' of possible rules for recognizing a flush we can now rule out those also shared by the non-examples—(3) and (5). One of the non-examples also contains 2, 3, 5, and 7, so we can rule out this feature. Both of the non-examples consist of cards of values less than 8, so we can rule out this feature. We are left with the 'rule' that a flush is probably a set of cards sharing a common suit.

The 'probably' is an important word here. Particularly in more complex and less clear-cut circumstances than a card game, when we are learning from examples—making general rules from particular instances—we must bear in mind the possibility that we may at some time in the future come across an example or non-example which forces us to modify our rules. In the case of a flush in poker this will not happen because we have succeeded in identifying a completely correct rule. In more complex aspects of life we must simply develop the best rules we can for the moment, and try to develop and improve them in the light of increasing experience.

We can translate the example of learning to recognize a flush into a form which the computer can handle fairly easily. First, we can give the computer the examples you have just been looking at:

```
is_a(example_1,flush).
consists_of(example_1(five,hearts,three,hearts,six,hearts,
  two,hearts,seven,hearts)).
is_a(example_2,flush).
consists_of(example_2(five,diamonds,two,diamonds,three,diamonds,
  four,diamonds,seven,diamonds)).
```

Translated back into English, these lines mean:

Example (1) is a flush;
Example (1) consists of five of hearts, three of hearts, six of
  hearts, two of hearts, seven of hearts;
etc.

The computer can now be told to find what examples (1) and (2) have in common. In other words, it must learn what the rule for recognizing a flush is without being told the rule itself. It must generalize from the particular examples given.

One way of doing this is to tell the computer to try replacing some of the specific details of the examples with *variables*. An illustration will make this clearer.

Let us assume (for the sake of clarity) that the computer tries the correct generalization first. In reality, it may well have come to this point after trying a number of incorrect rules. By replacing the names of the examples—(1) and (2)—with the variable (X) and by replacing the names of the suits—hearts and diamonds—with (Y), the computer can rewrite the statements above as follows:

```
is_a(X,flush):—
   consists_of(X(_,Y,_,Y,_,Y,_,Y,_,Y)).
```

This is translated as:

Any example (X) is a flush, *if*
that example (X) consists of any number at all, suit (Y),
any number at all, the same suit as the first card (Y),
any number at all, the same suit as the first card (Y),
any number at all, the same suit as the first card (Y),
any number at all, the same suit as the first one (Y).

Both examples (1) and (2) fit this one more general rule. Try it for yourself—substitute (1) for (X) and hearts for (Y), then substitute (2) for (X) and diamonds for (Y). *Both* examples fit the *one* rule.

In this way we can get the computer to learn. Instead of telling it which rules to use, we can just give it more general advice such as: 'To find a general rule underlying a number of specific examples, try substituting variables for the specific details and see if you come up with a rule which works.' This is just another way of saying what I would say to my companion: 'Find out those features which all the examples of flushes *share*.'

Having come up with a rule which does seem to work it is important to ensure that the rule does *not* work when applied to non-examples. Without this safeguard, incorrect rules may be generated. Whilst it may be true that some feature is shared by all the examples of flushes it may also be shared by the non-examples. For this reason it must be rejected. Again, try this for yourself—try substituting example (3) from page 119 for (X) in the rule above, and the suits in that example for the (Y). Since (Y) cannot be different suits (hearts and diamonds) at the same time, the rule above will not work for the non-example. For this reason, the rule can be retained.

We can equip the computer with various strategies for generalizing rules from examples. Much research work has gone into devising such generalization strategies which may involve such things as 'dropping conditions', 'adding alternatives', 'extending the reference', 'turning conjunction into disjunction', etc.

Don't worry about these rather offputting names. Names are often worse than the things they stand for. For example, we can easily think of an imposing title for what we did with the last card example. If we wanted to, we could say that we applied the *'variable substitution of constants' inductive inference* strategy. In any case, unless we want to get really involved in machine-learning programming, the details of such rules needn't concern us. Knowing something about machine-learning in general is, however, very important.

Complex learning

The card-game example in the last section represented a pretty simple sort of learning. There's quite a gap between learning what a 'flush' in a card game is and learning what rules an expert chess player is using—rules which may be so complicated that he or she may not even be able to tell you what they are!

Let's make things a little more complex.

> THINK!
> - Look at the list of people below.
> - Your task is to find out what characteristics tend to go with getting sunburnt easily.
> - In other words, you must try to complete the rule 'People who are likely to get sunburnt easily are those who..........'

Fred—red hair; blue eyes; skin type B. *Does* get sunburnt easily.
Carol—brown hair; blue eyes; skin type A. Does *not* sunburn easily.
Jackie—black hair; green eyes; skin type A. Does *not* sunburn easily.
George—red hair; brown eyes, skin type A. *Does* get sunburnt easily.
Ann—red hair; brown eyes; skin type C. Does *not* get sunburnt easily.
Mike—red hair; green eyes; skin type A. *Does* get sunburnt easily.
Mary—red hair; blue eyes; skin type C. Does *not* sunburn easily.
Jim—red hair; brown eyes; skin type B. *Does* get sunburnt easily.

There's one very simple answer—well done if you got it, but I don't think many people will, because in a way it's so obvious that it's extremely easy to overlook completely! This answer is that, according to the list, you're more likely to get easily sunburnt if you are male. In a way, this was a bit of a trick question designed to illustrate a very human, and altogether *uncomputer-like*, weakness—the ability to overlook the obvious! Now *please disregard* sex—*and try the task again to find the rule* 'Besides being male, people are more likely to get easily sunburnt if they have ....'

This is certainly a more complex rule-learning problem than the card-game example. If your answer is that people are more likely to get sunburnt easily if they have red hair, try again, because your answer is not complete (note that two people who do not get easily sunburnt do have red hair). Now here's my answer:

My rule is that people are more likely to get sunburnt easily if they have skin type B or skin type A and red hair.

If I did not beat you with that task, try this one. The reason for posing them will soon become clear.

> THINK!
> - This is a similar task. Below is a list of plant characteristics.
> - Your task is to discover which features tend to go with good fruit-production.

Root-length	*Root-texture*	*Root-growth*	*Fruit-production*
LONG	HAIRY	FAST	GOOD
MEDIUM	HAIRY	SLOW	GOOD
MEDIUM	SMOOTH	SLOW	POOR
LONG	SMOOTH	SLOW	POOR
LONG	SMOOTH	FAST	GOOD
SHORT	HAIRY	FAST	GOOD
SHORT	SMOOTH	FAST	POOR
MEDIUM	HAIRY	FAST	GOOD
LONG	SMOOTH	FAST	GOOD
SHORT	HAIRY	SLOW	POOR

The answer certainly involves a more complex rule than in the previous example. Here is my answer:

Good fruit-production is associated with (1) hairy medium or hairy long roots, (2) fast-growing, hairy, short roots, or (3) fast-growing, smooth, long roots.

Well done if you got it! But I shouldn't keep awake worrying if you didn't! It certainly requires quite a bit of effort and technique. Imagine how trivial a problem it is compared with the really complex problems tackled by human experts who have taken years to acquire their knowledge and skill.

The technique I'll describe now—to solve the last two problems—is central to a computer-learning program called 'ID3', devised by J. Ross Quinlan, which has been successfully used in what Professor Donald Michie, one of the most eminent researchers in artificial intelligence, considers to be genuine cases of a machine creating original knowledge (as opposed to machines helping people to make explicit and refine the perhaps hidden knowledge of human experts). The implications of this statement are indeed wide.

ID3 is a program which can automatically discover rules relating to complex problems. It has been used, amongst other things, to discover rules relating to a chess end-game 'of a level of complexity far beyond the power of the unaided chess master to codify adequately'. It should be able to help us with our sunburn and fruit-production problems!

Basically, ID3 works, as we did, from lists of characteristics of two types of phenomena—good fruit-bearing plants and poor fruit-bearing ones, easily sunburnt and not easily sunburnt people, winning chess moves and losing chess moves. The only thing is, ID3 can handle lists very much more complex than the ones we used. One series of chess experiments, for example, used lists of 49 different characteristics (we used three—hair-colour, eye-colour, skin type) and 715 different lists, representing 1 400 000 possible chess moves (we used eight for the sunburn example, ten for the plants example).

What ID3 does is to examine lots of examples—chess moves that worked, those that didn't; plants that did produce good fruit, those that didn't—and from these examples to 'distill' any underlying rules. The process is one of 'inductive inference'. This is also basically what you did—to infer rules from examples. However, how far did you use a technique similar to ID3? The best way to find out is for me to explain ID3's method.

ID3 creates a 'decision tree'. Figure 28 shows a decision tree (one of several possible decision trees) for the first problem we tackled. But how does ID3 produce decision trees? Let's take the first example we looked at. Here's the list again. This time, I've replaced the words 'does get sunburnt easily' with a tick ✓, and the words 'does *not* sunburn easily' with a cross ×, so that they're a little easier to spot when we use them in the decision tree.

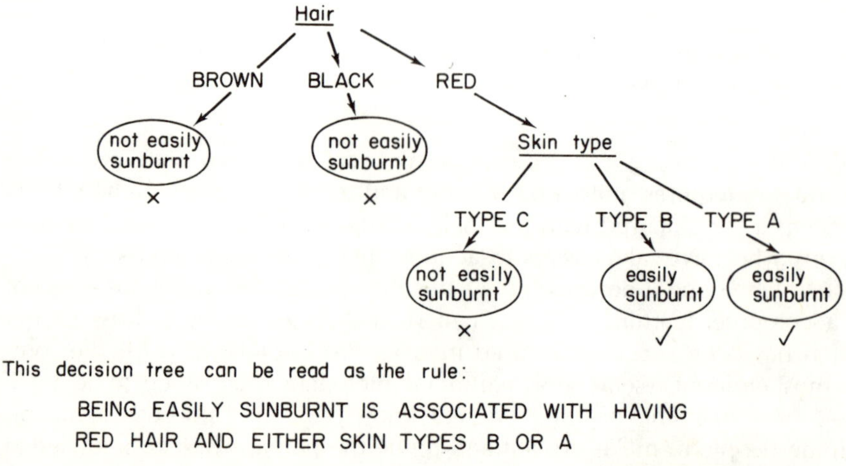

Figure 28. Decision tree for the 'sunburn' problem (1). (Based on Figure 15–2 in Quinlan, J. R. 1983. Learning efficient classification procedures and their application to chess end games. In *Machine Learning: An Artificial Intelligence Approach*, R. Michalski, J. Carbonell, and T. Mitchell, eds. Los Altos: Morgan Kaufman Publishers, p. 466)

	Hair	*Eyes*	*Skin type*
√ Fred	Red	Blue	B
× Carol	Brown	Blue	A
× Jackie	Black	Green	A
√ George	Red	Brown	A
× Ann	Red	Brown	C
√ Mike	Red	Green	A
× Mary	Red	Blue	C
√ Jim	Red	Brown	B

Now what we do is choose one of the features—hair, eyes, or skin—and draw a little 'tree' diagram for it as in Figure 29. We'll choose *hair*. We draw a line for each 'value' of hair—namely red, brown, and black. Next, we break up our original list according to each of the values—in other words, we group together all those with red hair, all those with brown, all those with black, as in Figure 30.

For any list in which we still have mixed ticks and crosses (as in the 'red hair' list above) we take another feature and subdivide the mixed list in the

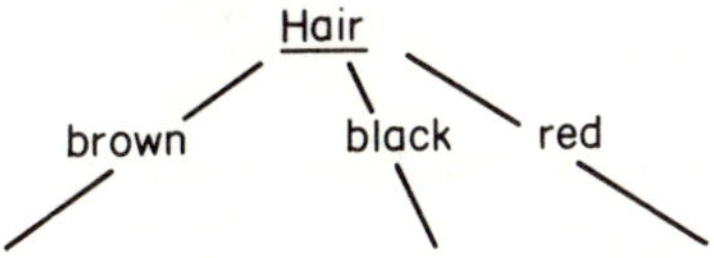

Figure 29. Decision tree for the 'sunburn' problem (2). (Based on Figure 15–1 in Quinlan, J. R. 1983. Learning efficient classification procedures and their application to chess end games. In *Machine Learning: An Artificial Intelligence Approach,* R. Michalski, J. Carbonell, and T. Mitchell, eds. Los Altos: Morgan Kaufmann Publishers, p. 466)

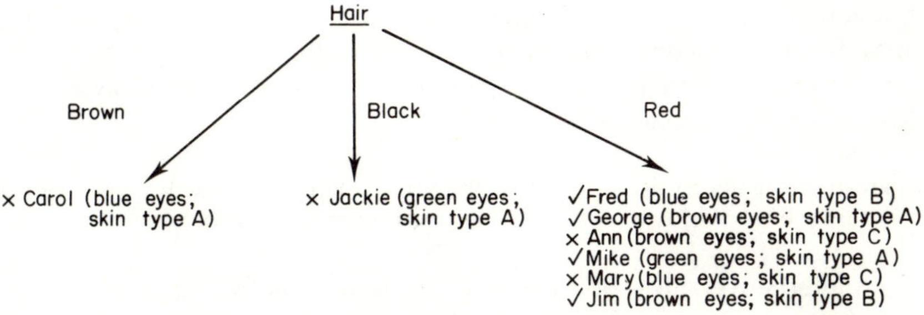

Figure 30. Decision tree for the 'sunburn' problem (3). (Based on Figure 15–1 in Quinlan, J. R. 1983. Learning efficient classification procedures and their application to chess end games. In *Machine Learning: An Artificial Intelligence Approach,* R. Michalski, J. Carbonell, and T. Mitchell, eds. Los Altos: Morgan Kaufman Publishers, p. 466)

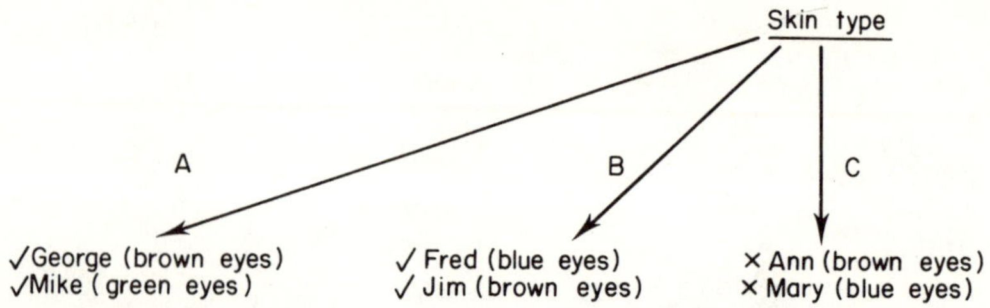

Figure 31. Decision tree for the 'sunburn' problem (4). (Based on Figure 15–2 in Quinlan, J. R. 1983. Learning efficient classification procedures and their application to chess end games. In *Machine Learning: An Artificial Intelligence Approach,* R. Michalski, J. Carbonell, and T. Mitchell, eds. Los Altos: Morgan Kaufman Publishers, p. 466)

same way. If we take *skin type* as our next feature, the decision tree will grow as in Figure 31.

Now that you have no lists left which contain mixed ticks and crosses, your decision tree is complete. This tree may be read as the rule:

> Getting easily sunburnt (✓) is associated with skin types A and B but only in red-haired people.

A complementary rule derived from the tree is:

> *Not* getting sunburnt easily (×) is associated with having brown or black hair, or having red hair and skin type C.

You may have thought that in this example things might have turned out differently if we had chosen features in a different order. We could have started with *skin type* instead of *hair*, or chosen *colour of eyes* for our second feature, etc. You would have been right! The decision tree would have looked much different—maybe much bigger. The rules we would have ended up with may have been very different but equally valid.

Figure 32 shows an alternative decision tree for the 'sunburn' problem. This decision tree may be read as the rule:

People are more likely to get sunburnt if they have skin type B with brown or blue eyes; or if they have skin type A with either green eyes and red hair, or brown eyes.

This is an alternative rule—but it works just as well as that derived from our first decision tree. There is one thing, though—it's just a bit more complex than our first rule. Let's compare the two:

First rule (from the first decision tree)
Sunburnt if (1) red hair *and* skin type A; *or*
(2) red hair *and* skin type B.

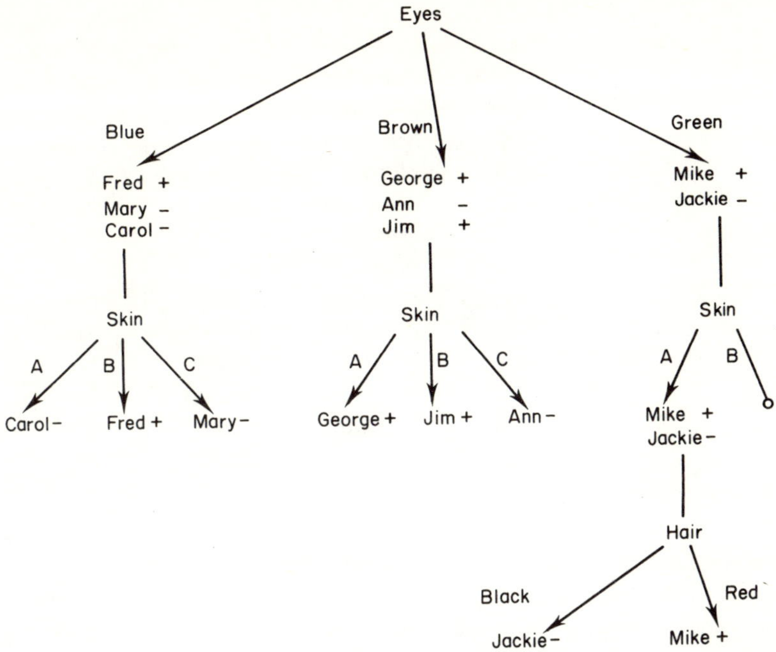

Figure 32. An alternative decision tree for the 'sunburn' problem (1). (Based on Figure 15–2 in Quinlan, J. R. 1983. Learning efficient classification procedures and their application to chess end games. In *Machine Learning: An Artificial Intelligence Approach*, R. Michalski, J. Carbonell, and T. Mitchell, eds. Los Altos: Morgan Kaufman Publishers, p. 466)

Second rule (from second decision tree)

Sunburnt if (1) skin type B *and* brown eyes; *or*
(2) skin type B *and* blue eyes; *or*
(3) skin type A *and* green eyes *and* red hair; *or*
(4) skin type A *and* brown eyes.

The second rule is not as economical as the first (a simple count of the number of '*or*'s and '*and*'s shows this). Particularly for complex problems, there is much virtue in discovering economical, minimal rules rather than very complex, lengthy ones. But how can we select economical rules?

THINK!
- Look at the lists of features below.
- Imagine that you are to discover a rule showing what features tend to go with getting good examination results at school.
- Don't produce a decision tree—simply think *which features you would choose first and second* for this process and why.

Exam performance	*Shoe-size*	*First letter of name*	*Intel-ligence*	*Favourite colour*	*Works hard*
Good	8	G	Medium	Red	Yes
Good	4	A	High	Brown	Yes
Poor	7	P	Medium	Red	No
Poor	9	G	Low	Blue	Yes
Good	2	U	High	Red	No
Poor	3	H	Medium	Pink	No
Poor	7	S	Low	Red	Yes
Good	9	I	High	Blue	No
Good	3	W	Medium	Brown	Yes
Poor	6	B	Medium	Pink	No
etc.	etc.	etc.	etc.	etc.	etc.

No prizes for this one. Unless I have misjudged you, the first two features you would choose are (in either order) intelligence and hard-working habits. I can't imagine many people opting for shoe-size or first letter of the name!

The point is—what you found very easy and obvious would be neither easy nor obvious to a computer. What you have that it hasn't is wide-ranging common sense and knowledge of the world and how people work.

ID3 tries to get over this problem of deciding which features to choose first in the interests of producing the most economical, minimum set of rules possible. However, it does so mathematically rather than 'common sensibly'. Are the two so different? Let me explain ID3's way, and you decide whether it has any of the flavour of what we call 'common sense'.

The first choice, *skin type*, leaves us four people yet to sort out into separate tick (✓) and cross (×) categories from a total sample of eight people—call this 4/8 yet to do.

The second choice, *hair-colour*, leaves us with six people yet to sort out—call it 6/8 yet to do.

The third choice, *eye-colour*, leaves us with all eight yet to sort out—there are *no* lists containing either all ticks or all crosses.

On this basis, we would do best to apply the first choice, *skin type*, as our first feature. A similar process can calculate the best choice for the second feature, and so on. This is rather like the process of calculating the best move in the 'number boxes' game discussed in Chapter 5.

If we chose skin type, the first branches of our decision tree would look like this:

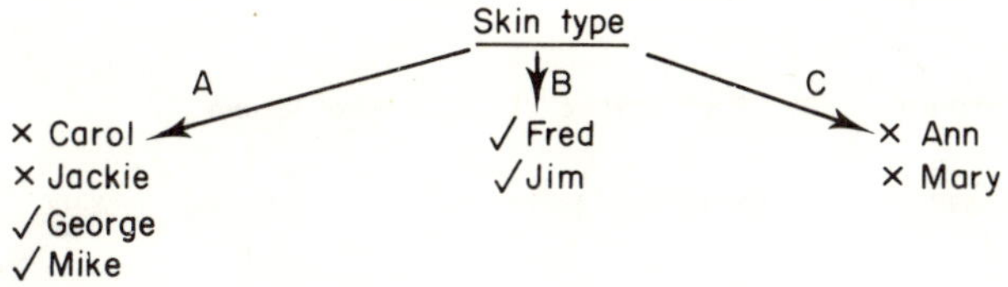

Choosing hair-colour gives us this alternative:

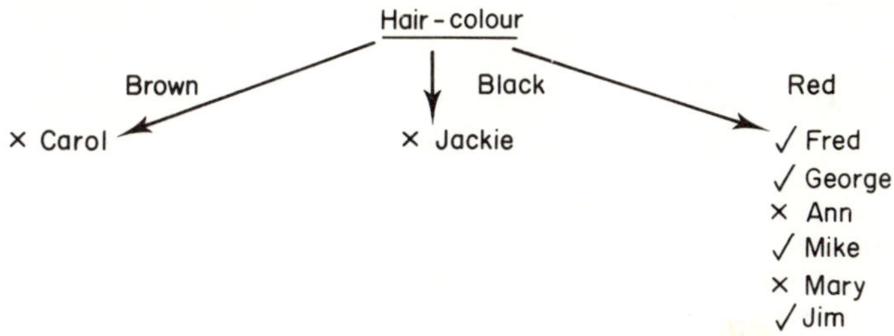

Whereas eye-colour would give us:

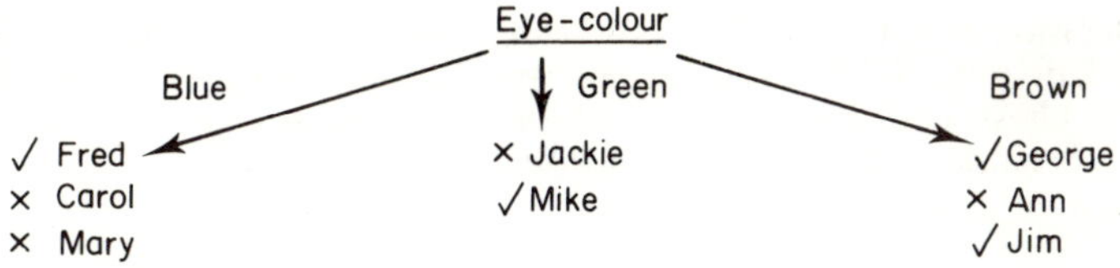

Figure 33. An alternative decision tree for the 'sunburn' problem (2). (Based on Figure 15–1 in Quinlan, J. R. 1983. Learning efficient classification procedures and their application to chess end games. In *Machine Learning: An Artifical Intelligence Approach*, R. Michalski, J. Carbonell, and T. Mitchell, eds. Los Altos: Morgan Kaufman Publishers, p. 466)

THINK!
- Turn back to the 'fruit-production' problem you tackled before.
- Produce a decision tree for it. The answer I've given in Figure 34 is based on choosing *root-length* first, followed by *root-texture* then *root-growth*.
- Feel free to make different choices—but your tree will differ from mine. You may come up with a more economical rule than the one my tree produces.

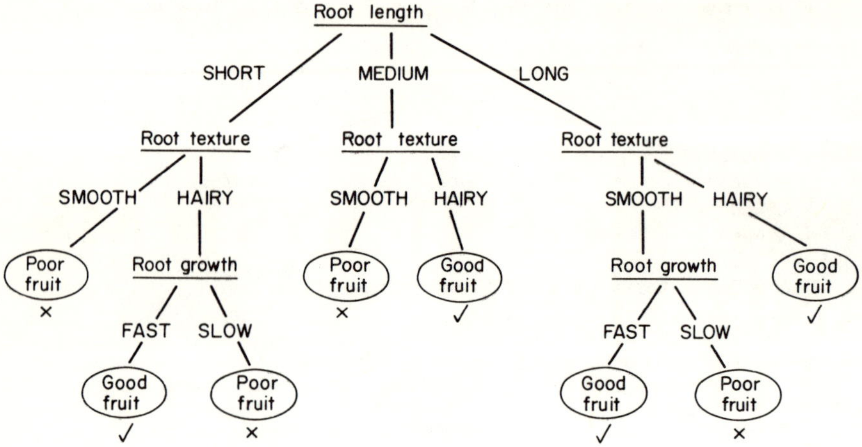

This decision tree can be read as the rule
GOOD FRUIT PRODUCTION IS ASSOCIATED WITH
(a) MEDIUM HAIRY OR LONG HAIRY ROOTS; or
(b) LONG, SMOOTH, FAST-GROWING ROOTS; or
(c) SHORT HAIRY FAST-GROWING ROOTS

Figure 34. Decision tree for the 'fruit-producing plants' problem. (Based on Figure 15–2 in Quinlan, J. R. 1983. Learning efficient classification procedures and their application to chess end games. In *Machine Learning: An Artificial intelligence Approach*, R. Michalski, J. Carbonell, and T. Mitchell, eds. Los Altos: Morgan Kaufman Publishers, p. 466)

Machine learning by induction is a particularly important and exciting area of machine intelligence with great potential in the development of expert systems. It has already produced very exciting and important results. One system which makes use of it—EXPERT-EASE—is discussed in Chapter 12.

Part C:

INTELLIGENT SYSTEMS AND THEIR DEVELOPMENT

CHAPTER 12

WORKING SYSTEMS

So much for some of the processes and strategies available to the designers of intelligent computer systems. Now let's take a look at some of the systems that have been created. They have been chosen to show a range of different types of application and to illustrate some of the important concepts introduced in Parts A and B.

STRIPS: means–ends analysis

STRIPS (Stanford Research Institute Problem Solver), devised by Richard Fikes and Nils Nilsson, can solve a wide range of problems using means–ends analysis, discussed in Chapter 6. One example of its use has been as the problem-solving part of a simple robot—SHAKEY.

SHAKEY is a device on wheels which can move around and push objects. 'He' works in a series of rooms connected by doors, solving problems which require boxes to be moved from room to room. Before he can move a box, SHAKEY must ensure that he is next to it. Before he can go from one room to another, he must ensure that there is a door connecting the two rooms, and that the door is open. STRIPS produces a plan which SHAKEY puts into action to solve a problem. Although SHAKEY cannot physically climb on boxes and open windows, STRIPS has also been required to produce *plans* for tasks such as:

'Go from room 5 to room 1 via room 4, then open window 1 by climbing on a box moved from room 4.'

Just as in Chapter 6 we used operators ('fly' and 'bus') to plan our journey from Manchester to Murcia, STRIPS uses *operators* with *preconditions*. STRIPS operators also have an *add* list, and a *delete* list. Examples of relevant operators are 'climb_on' and 'move_to', shown in Figure 35.

The original STRIPS system solved every problem afresh. In other words, when presented with a problem that it had solved previously, it would not be able to use its experience by remembering how it had done it the first time.

STRIPS was therefore developed so that it could *learn*. Imagine the following problem, which can be solved using the two operators shown in Figure 35:

'Move to box 1 in room 2 and climb onto it.'

Operator	Preconditions	Add	Delete
climb_on(box(X))	is_next_to(robot,box(X)) is_on(robot,floor)	is_on(robot,box(X))	is_on(robot,floor) is_next_to(robot, box(X))
move_to(box(X))	is_on(robot,floor) is_in(robot,room(Y)) is_in(box(X),room(Y))	is_next to(robot, box(X))	is next to (robot,_)

Figure 35. 'SHAKEY' operators. (Adapted from Open University (1978). D303, *Cognitive Psychology*, Block 4, Units 26/27, p. 44)

The designers of STRIPS developed the idea of 'super-operators' (which they called MACROPS). These represent combinations of operators which have been used in the past to solve complex problems. The preconditions of each of these operators are also recorded, as are details of the 'add' list.

All this information is stored in 'triangle tables' as shown in Figure 36. In this figure all the boxes to the left of an operator contain that operator's preconditions. All boxes underneath an operator contain that operator's 'add' list. In other words, operator 1, 'move_to(box(1))', has the preconditions 'is_on(robot,floor)', 'is_in(robot,room(2))', and 'is_in(box(1),room(2))', and when applied will add to the computer's memory 'is_next_to(robot,box(1))'.

Operator 2, 'climb_on(box(1))', can only be applied if the preconditions 'is_on(robot,floor)' and 'is_next_to(robot,box(1))' are met. When applied, operator 2 will add 'is_on(robot,box(1))' to the computer's memory.

The whole triangle of boxes represents the *combination* of the two operators:

move_to(box(1)),
climb_on(box(1)).

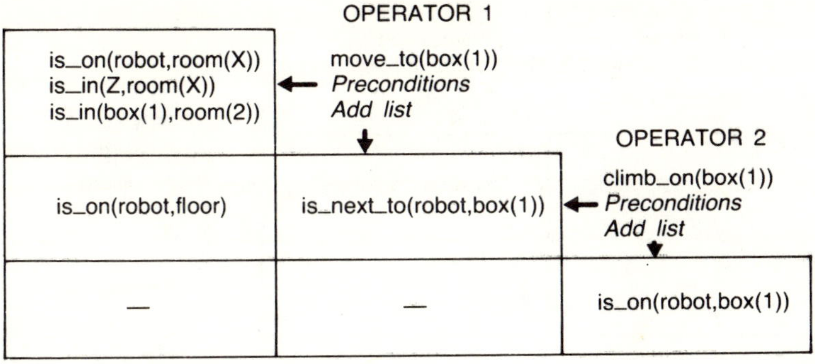

Figure 36. An example of a triangle table. (Based on Figure D5–1 in STRIPS, by Anne Gardner. In *The Handbook of Artificial Intelligence*, vol. 1, Avron Barr and Edward A. Feigenbaum (eds), Los Altos: William Kaufmann, Inc., 1981, pp. 128–34)

with the preconditions:

is_on(robot,floor),
is_in(robot,room(2)),
is_in(box(1),room(2)).

and an add list:

is_on(robot,box(1)).

What STRIPS does to make this combination more useful is to generalize it. This means that instead of having a super-plan which can be applied if SHAKEY has to climb on box 1 in room 2, STRIPS has a super-plan which can be applied if SHAKEY has to climb on any box in any room they are both in. This is done by substituting variables for constants. Compare Figure 36 with Figure 37, which represents a generalized version of the same super-operator shown in Figure 36. This MACROP is:

move_to(Z),
climb_on(Z).

with the preconditions:

is_on(robot,floor),
is_in(robot,room(X)),
is_in(Z,room(X)).

and an add list:

is_on(robot,Z).

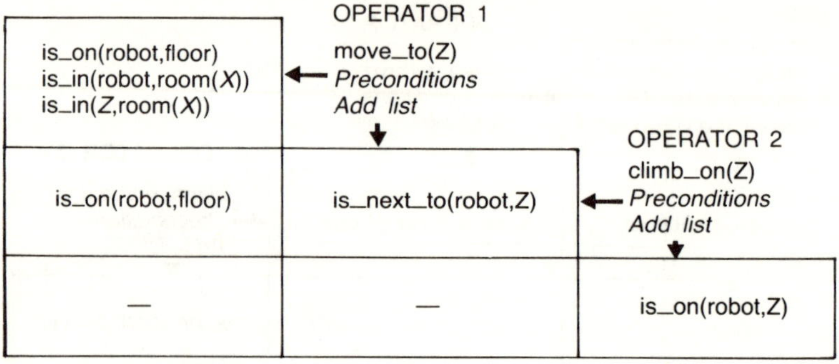

Figure 37. A generalized version of the triangle table shown in Figure 36. (Based on Figure D5–1 in STRIPS, by Anne Gardner. In *The Handbook of Artificial Intelligence*, vol. 1, Avron Barr and Edward A. Feigenbaum (eds), Los Altos: William Kaufmann, Inc., 1981, pp. 128–34)

In other words, the robot can move to and climb on any object in any room, provided that it is on the floor and that both are in the same room. This MACROP will be considered whenever SHAKEY wants to achieve a goal which has the form 'is_on(robot,Z)'—in other words, whenever he wishes to climb onto anything. It represents a remembered combination of operators—the memory of how a similar problem was tackled in the past. SHAKEY can use this experience, rather than re-solving the problem from scratch. MACROPS can, of course, be much more complex than this, and consist of many more boxes arranged in a triangular form. In this way, complex plans can be remembered in a generalized form, and applied when the same problem, or a different problem with the same characteristics, crops up.

A different program—PLANEX—is used to make the most of these MACROPS by applying them in the most efficient way. PLANEX does not just apply each operator making up a super-operator in turn. In some circumstances, for example, it may be more efficient to miss out certain steps of the plan, or for some reason a certain step may not succeed in achieving its expected effect. To overcome such problems, PLANEX reads the triangle tables, starting at the bottom, and applies the first operator it can find with all its preconditions satisfied. Having done so, it starts again at the bottom of the tables.

ABSTRIPS: planning at separate levels

Even though the means–ends analysis used by STRIPS is far removed from the 'brute'force' approaches discussed in Chapter 5, in complex problem-solving there is a need to further reduce the 'combinatorial explosion'—in other words, the vast range of *possible* paths through the problem state space.

ABSTRIPS was developed by Earl Sacerdoti as a modification of STRIPS, designed to reduce the processing required in problem-solving. It is based on the idea of hierarchical planning—breaking down a problem into different layers, or levels of abstraction, using the principle described in Chapter 7.

The ideas behind ABSTRIPS can best be seen if we move away from the rather narrow world of SHAKEY. Imagine again our proposed holiday in Murcia, and that our computer system is going to plan things in rather more detail than before. As we saw in Chapter 7, key issues to decide are when you will be going, where to, and for how long. Having decided this, you would then probably go on to consider how you would travel—bus, taxi, train, plane.

Importantly, other things would not be considered at this preliminary stage. You would not be concerned to plan the small details of your journey until you had firmly decided where you were going. Some things you couldn't do—you couldn't, for example, book your plane tickets until you knew exactly when you were to travel. Other things you wouldn't worry about, because they could easily be solved. You wouldn't look in your pocket, find you had no change, and so decide in advance not to plan on taking a taxi. The reason is that you know that on the day on which you travel you will make sure you have some change. Similarly, you will not be particularly concerned months before to look up the taxi's telephone number. This can be left until the day or possibly the night before if you are booking in advance.

In other words, you would be making plans at different 'layers', not getting bogged down in small details when deciding the important things. Only when things were planned at the 'top' layer of most important decisions—where and when you were going—would you need to plan at a lower level—actually booking tickets, sending for passports, choosing a particular flight. Only when this second layer of plans was satisfactorily completed would you need to plan the third-level details—remembering to lock the door, cancel the milk, make sure you had the taxi fare, actually telephone for a taxi, etc.

ABSTRIPS plans in 'layers' like this, avoiding getting held up in details at an early stage. If a plan doesn't work at the top level this will be noticed before time and effort has been expended at lower levels. ABSTRIPS checks through all the details at each level before going on to the next. In other words, the more detailed lower levels will only be considered and processed if higher levels have been checked and will work.

The procedure used by ABSTRIPS is to examine all the systems' *operators* (like 'fly', 'bus', 'climb_on', 'move_to') and to give each of their preconditions criticality values. The highest criticality values are given to preconditions which cannot be changed by any of the system's other operators.

For example, a precondition for flying from place (A) is that place (A) has an airport. There are no operators which can change this—if you are in Bolton, there is no operator which can 'give' Bolton an airport! This precondition would be assigned the maximum score—call it 6. All the other preconditions are given scores, the score being lower, the more easily the precondition can be changed. In this way the lowest score would be given to preconditions relating to where the traveller is—operators *can* be applied to change this. Quite a complex formula is used for deciding on these criticality values. This is done by human and machine means. First, the programmer makes a partial listing, in order of priorities, of the operator preconditions. Then the system works out in more detail the criticality values, based on an analysis of the levels of difficulty with which each precondition could be changed.

One series of preconditions may all score 6, another series 5, and so on down to 1. All the preconditions with the same number share the same level of abstraction (see Figure 38). What ABSTRIPS does is to process the highest level first. In other words, when forming a plan by selecting operators which, it is hoped, will eventually solve the problem, all preconditions with criticality

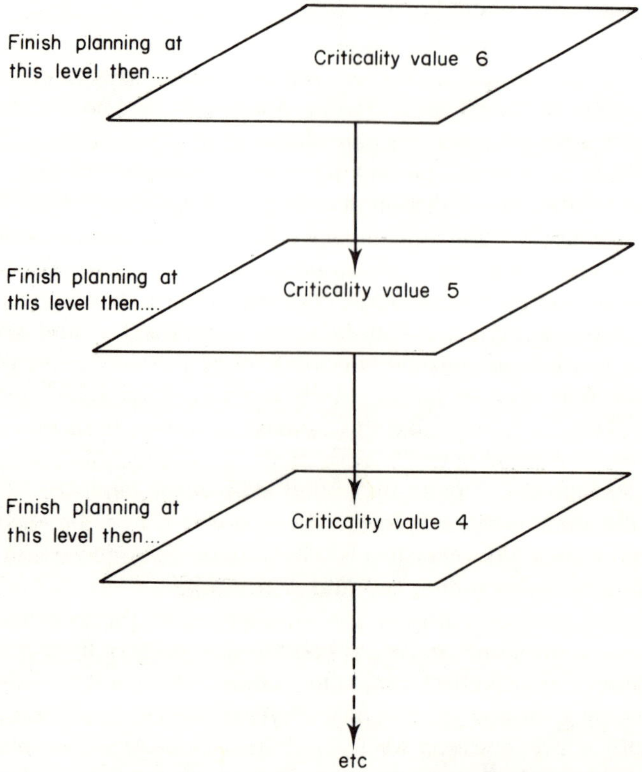

Figure 38. Planning at separate levels

values below those of the level being processed are ignored. When planning at the top level (criticality value 6) all preconditions with criticality values of less than 6 are ignored. Only when the problem has been 'solved' (that is, no problem has been encountered) at this top level does ABSTRIPS move onto the next. At *this* level, all preconditions with a value of less than 5 will be ignored.

This approach is useful in recognizing dead ends early on and saving wasted effort. If something goes wrong at a lower level, the system backtracks to its immediately higher one, and the 'junction' leading to where the problem was found is blocked off. The system then continues searching for a solution in that higher level. ABSTRIPS contains some further procedures to improve its performance by ensuring that the shortest solution route is found and by delaying certain choices until sufficient information is available as processing continues.

The performance of STRIPS and ABSTRIPS in devising plans for SHAKEY has been compared over a number of problems. For one of the longest problems, STRIPS found a solution using 30 minutes of computer time. ABSTRIPS solved the same problem using just under $5\frac{1}{2}$ minutes!

MOLGEN: planning at interacting levels

ABSTRIPS reduced the amount of processing required in finding solutions to a problem by breaking the problem down into different levels of 'criticality'—leaving things that could be changed or achieved relatively easily until last. However, essentially, processing at one level was completed before processing at a lower one was started. In this way the different layers of planning were largely independent from one another. Information on *why* planning at a lower level had failed, if it had, was not passed up to and used by the upper layers.

MOLGEN is a system designed by M. J. Stefik at Stanford University to solve problems again by working at different levels but allowing interaction between the different levels. In other words, the different levels can 'talk to' each other. If a problem is found at a lower, more detailed level, information about the problem is passed up to higher levels which can take them into account in their own planning. The complete plan is not formed by simply completing processing at the top level, then completing processing at the next level, then the next level, and so on. Rather, planning goes on with the different levels communicating and interacting with one another.

The system is actually a very specialized one, designed to help scientists design experiments in molecular genetics. The sort of problem MOLGEN is best at solving is, for example, helping to design an experiment to produce a culture of bacteria to produce insulin. However, for the sake of clarity, a much simplified example is given here. This also relates to a problem a little more familiar to most readers than molecular genetics. It must be remembered, though, that it is used only to show the sort of procedures used by MOLGEN so that they can be compared and contrasted with the other systems described in this book. The real MOLGEN is used to solve rather more specialized scientific problems.

MOLGEN has three levels of problem-solving processes (Figure 39). The top one is concerned with strategic planning—proposing new planning steps, saying when pending steps should be put into action, and undoing planning decisions found not to be feasible. The middle level is concerned with more detailed planning decisions involved in selecting operations from the bottom level. It fits these operations together as it refines and fills in the details of the overall plan. The bottom level is concerned with the actual activities that can be orchestrated by the higher levels in such a way that the problem can be solved.

The following example shows the sort of problem-solving strategy used by MOLGEN. A MOLGEN-type system, then, when given the problem of planning a journey from Bolton to Amiens in northern France, might propose the basic plan:

BOLTON → DOVER → CALAIS → AMIENS
drive ferry drive

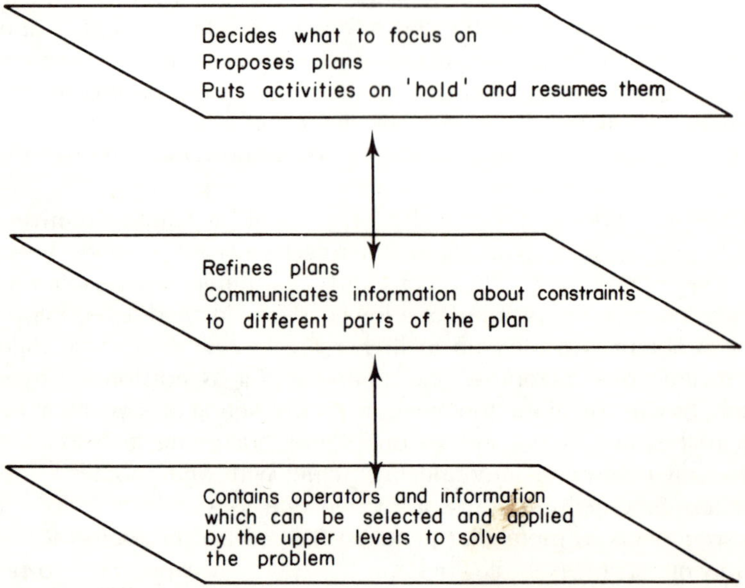

Figure 39. Planning at interacting levels. (Adapted from Stefik, 1980)

The top level might decide to focus on refining the first 'drive' operator—in other words, to examine it in more detail and instruct the second level to carry out this refinement. This first 'drive' part of the plan would require a number of more detailed steps, one of which might be making sure all necessary documents had been obtained. At this stage, the second level might find that certain documents would be required, but at this stage full details of precisely which documents might not be available. In such circumstances, the top level can decide to suspend work on refining the 'drive' operator—to put it on 'hold'—and decide to focus on, say, refining the 'ferry' operator. However, before moving on to do so it would make a note that certain documents (as yet unspecified) will be necessary for the first 'drive' stage. When the *ferry* operator is being examined in detail at the bottom level the requirement to have a valid passport would emerge. The top-level processes would now step in, realizing that this is in fact one of the documents that will be required for the 'necessary documents' part of the plan. It would tell the second level to pass the relevant information—that one of the necessary documents must be a valid passport—back to this earlier part of the plan. A similar process will occur when the second 'drive' part of the plan—this time relating to driving in France—is examined in detail. As the requirement to have a valid visitor's driving licence emerges this information can be passed back to the 'necessary documents' stage. In this way, different parts of the plan can gradually be refined, certain details being left open until more information becomes available. The top level would search for another task to get on with. If none could be found, it would examine all the operations on 'hold' to see if sufficient information had been obtained since they were put on hold to resume one of them. If not, then the top level might undo previously made decisions and suggest a new plan.

Basically, the control is hierarchical. Although there is two-way communication between levels, the top level is firmly in charge of what goes on.

SAVOIR: dealing with probabilities

It is possible to buy the frameworks—or 'shells'—of expert systems in ready-made form. In other words, the programming of the system to draw conclusions from evidence has been done for you. What you must do is provide the knowledge. In this way, an expert-system shell can be applied to a wide range of different problems and areas of knowledge. Whilst these differ, the basic mechanism remains the same. A number of expert-system shells are available, ranging from relatively simple ones for home computers to highly sophisticated ones designed for large computers owned by industrial and commercial organizations.

SAVOIR is a sophisticated expert-system shell designed by ISI for commercial and industrial markets. It includes a good many features, only some of which are available on some of the cheaper systems suitable for smaller computers. It is a rule-based system—in other words, its knowledge is represented in the form of rules which state that IF certain conditions are met THEN certain conclusions follow. We looked at this sort of representation in Chapter 2. So SAVOIR can handle knowledge such as that we met in the 'car faults' system in Chapter 8:

IF the engine will not start
and the lights are dim THEN the battery is likely to be flat

SAVOIR allows the user of the system to answer questions with varying levels of certainty. Some statements can be clear-cut. For example, if the system needs to know whether the car will start there are only two possible answers—'yes' or 'no'. With other questions the answers may vary. When asked if the lights are dim, the user may want to say 'a bit' or 'very much' rather than just 'yes' or 'no'. When asked whether there is a short circuit, the user may not be sure, and wish to reply 'it's possible', 'very likely', or 'highly unlikely', depending on a variety of factors such as whether he or she had checked the wiring only the day before or had constantly had trouble with the wiring over recent months! SAVOIR allows users to reply in this 'fuzzy' way, as well as giving clear-cut 'yes' and 'no' answers where appropriate.

The system combines these different answers with an estimation of the relative strength with which each conclusion depends on them, and calculates the probability, or likelihood, of the conclusions it comes to. It does this by using Bayesian updating—discussed in detail in Chapter 8.

For example, we could build into SAVOIR the knowledge represented in Figure 40. Imagine that disease A has prior odds of 0.3 of occurring in the population as a whole—according to the expert who is giving us the knowledge. The likelihood that a particular patient has the disease may be inferred from whether or not he or she has disorder B *and* whether or not he or she complains of symptom E. Disorder B can be inferred from whether or not the patient complains of symptoms C and D.

However, whether or not the patient has disorder B, or symptoms C, D, or E, may not be clear-cut. There may be a 65% likelihood that he or she has disorder B. The patient may definitely have symptom C, may be unsure, or just have slight signs of it. All this uncertainty is handled by SAVOIR in the following way.

When asked by the system whether a particular symptom is present the user can reply using any number from +5 to −5 to indicate a range of certainty from definite ('true' in Figure 40) to definitely not ('false'). Let's take symptom C, for example. The prior odds of anyone having symptom C is, according to our expert, 0.6—in other words, it is a pretty common symptom. If the user of the system replies 5 (i.e. the symptom is definitely present), then

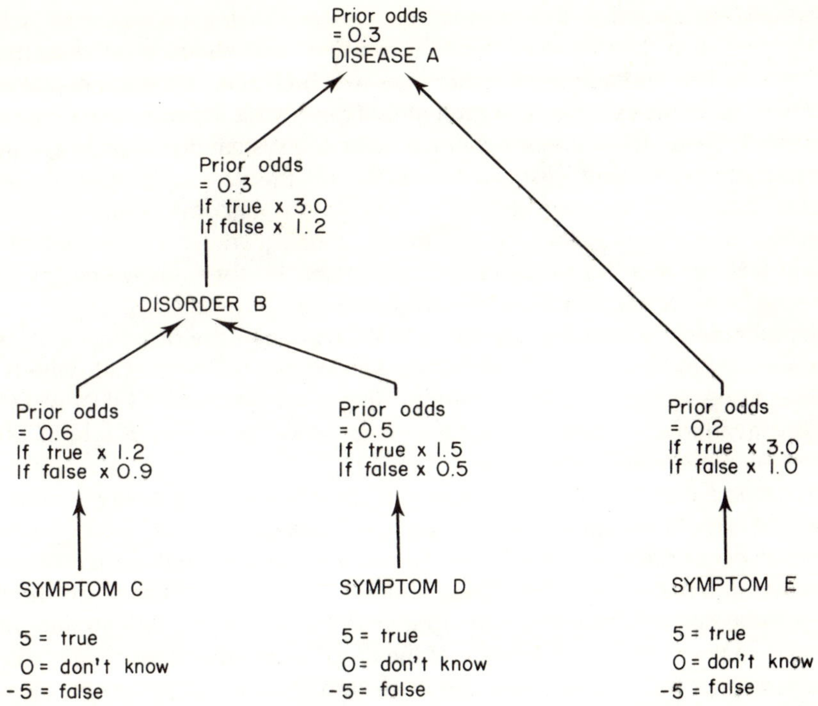

Figure 40. Dealing with probabilities in SAVOIR

these prior odds are multiplied by 1.2 (see Figure 40). This 1.2 figure is the logical sufficiency factor (discussed in Chapter 8), and has been provided by the expert on the basis of his or her knowledge and experience. If the user replies −5 (i.e. the symptom is definitely not present), then the prior odds are multiplied by 0.9—the logical necessity factor, derived in the same way from our expert. If the user replies with a number between +5 and −5, the system automatically calculates what number between 1.2 and 0.9 by which to multiply the prior probability, as explained in Chapter 8. The end-product may be a diagnosis that the patient may have disease A with a probability of 0.65, or a probability of 0.54, etc.

In Chapter 2 we met backward- and forward-chaining. SAVOIR uses both in a very effective way. Backward-chaining is used to ask questions relevant to the hypothesis it is working on at the time. If it thinks a particular diagnosis is likely it asks questions specifically related to proving or disproving that diagnosis. However, as questions are answered, and as the user may volunteer information he or she considers important, SAVOIR uses forward-chaining to take account of this evidence, which can be used to revise beliefs and, if necessary, to abandon the current line of questioning in favour of a more productive one. In this way, both 'top-down' and 'bottom-up' information can be used in interaction. Another way in which relevant 'bottom-up' information can be made use of immediately is by using actions—also known as 'demons'. These allow the designers to make the system do something (change the line of questioning, display a message on the screen, investigate a particular idea, etc.) as soon as certain conditions become fulfilled. In other words, the system can be put 'on alert', waiting until a certain set of preconditions are met (for example, by certain of the user's answers), at which time the actions are automatically triggered off.

A particularly interesting use of SAVOIR has been the development by ICI of a series of expert systems collectively known as COUNSELLOR, which give arable farmers expert advice on controlling crop diseases. COUNSELLOR is widely available to users by means of a viewdata interface which allows access via the existing telephone network.

By asking the farmer questions about the type of crop being grown, soil type and conditions, etc. COUNSELLOR will assess the risk of various diseases and consequent possible yield-loss. The system can then recommend the most effective treatment, with detailed costings. It can also work out, on the basis of current costs of buying and applying the recommended treatments in relation to the current selling price of the crop, a detailed cost-benefit analysis. It can do this for a variety of alternative programmes of treatment, allowing the farmer to select the one he or she thinks most effective. Recommended treatments are not limited to those produced by ICI. Competitors' products are also included! COUNSELLOR can also be questioned as to why it has asked a particular question or made a particular recommendation: in effect, to justify its reasoning to the farmer.

SAVOIR has also been used in a variety of applications, including process control, engineering design, the diagnosis of faults, choosing insulation for pipework appropriate to particular chemical environments, and predicting cracking in pipework due to stress and corrosion. British Aerospace is also using SAVOIR to construct an expert system to help airframe designers in a military aircraft and guided-missile establishment. The system—FEASA (Finite Element Analysis Specification Aid)—when finished will be used in combination with an existing set of highly complex computer programs. Together, these systems will help the designers in the process of analysing complex problems of mechanical stress.

HEARSAY: blackboard techniques

Paradoxically, in order to perform a task that we can all do without even thinking—understanding speech—very sophisticated and complex systems have been developed. One of these systems is HEARSAY.

One of the best ways of coming to realize just how complicated is 'ordinary' understanding of what we hear, is to try to devise a computer system capable of performing the task.

> THINK! • How would you distinguish between the following if you heard them spoken and could not see the different spellings?
>
> I scream/ice cream
> ate/8
>
> and between the following if you heard them without the explanations in brackets?
>
> Kicking legs can be painful (= legs which kick).
> Kicking legs can be painful (= it is painful to kick them).

The answer is that the context in which you heard these words or expressions—the words coming before and after, and the situation in which you heard them (who was speaking, what were they talking about at the time, etc.)—would have given you clues. Your knowledge of words, grammar, and meanings would do the rest. We looked at the effects of context in Chapter 9.

Linguists generally agree that when we understand speech or written language we do so partly on the basis of what we see or hear—what is actually said or written—and partly on the basis of what we expect to see or hear. In other words, as well as hearing or reading sounds or written symbols we are also constantly using expectations about what is likely to be being said or written in the particular context in which we find ourselves. These expectations allow us to ignore, unless the message is very unclear, all the possible ambiguities of interpretation—the same sounds representing different possible words, the same words having different possible meanings, etc.

In this way we use both 'bottom-up' processing (in the case of speech understanding, perceiving the sounds being spoken) and 'top-down' processing (interpreting these perceptions in the light of our expectations—constantly forming hypotheses about the meanings of what is being said, and which of various possible interpretations of sounds, words, and meanings is the most likely).

At any time, when listening to spoken language we will have a number of perhaps competing hypothesis in our minds at the same time, some being confirmed, others being rejected, and others being formed, as time goes on. These hypotheses will constantly be 'played off' against incoming aural perceptions, as 'top-down' and 'bottom-up' processing interact in our attempts to make sense of what we are hearing.

HEARSAY was designed at Carnegie–Mellon University to try to cope with this complex task. The basic approach described here has also been used in systems built to perform a wide variety of tasks in addition to speech-understanding, including problem-solving in crystallography, vision, signal-interpretation, and psychological modelling.

HEARSAY uses a 'blackboard', divided into a number of different levels relating to parts of words, whole words, sequences of words, and phrases (see Figure 41). In addition to the blackboard are a number of independent 'knowledge sources' with specialized knowledge about sounds, syllables, words, word-sequences, and phrases.

The knowledge sources in charge of sounds interpret the sounds in terms of possible syllable classes. They 'write' these hypotheses on the blackboard, which is 'read' by other knowledge sources. The knowledge sources in charge of words will read the hypotheses on the blackboard and in turn hypothesize likely words. These hypotheses are also written on the blackboard at the 'words' level. Here they can be read by the knowledge sources in charge of word-sequences, which, in turn, will write hypothesized word-sequences on the blackboard. These hypotheses will be interpreted by the 'phrase' knowledge sources to produce a hypothesized phrase which, hopefully, will represent a successful interpretation of all of the original sounds.

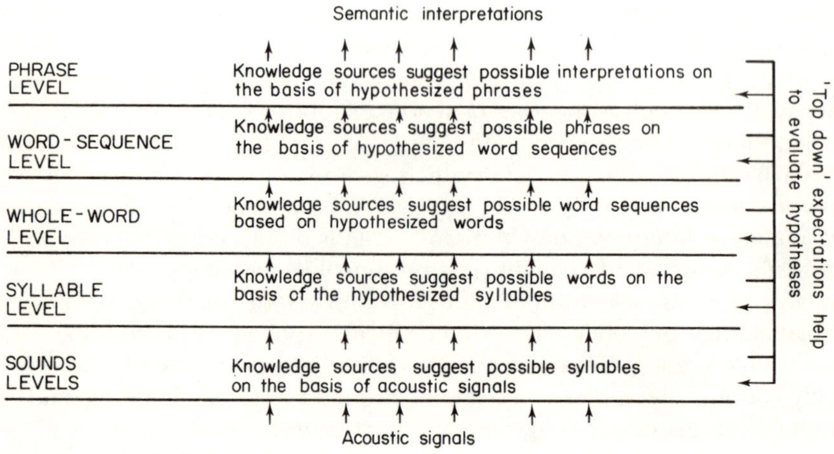

Figure 41. The concept of a 'blackboard' and different knowledge sources, based on HEARSAY II. (Hayes-Roth, Waterman, and Lenat, *Building Expert Systems*, © 1983, Reading, Mass.: Addison-Wesley, p. 118, Fig. 4.8 (adapted). Reprinted with permission)

The knowledge sources are independent of one another and can respond to hypotheses (i.e. modify existing hypotheses or create new ones) on their own initiative. This 'initiative' is a triggering procedure activated by the characteristics of a particular hypothesis. In other words, each knowledge source will look for a particular form or pattern of hypothesis, indicating that it is suitable for treatment by that particular knowledge source.

When more than one knowledge source is activated at the same time another special knowledge source called a *scheduler* keeps control of the order in which the other knowledge sources go ahead with their interpretations. Its decisions are based on a scoring system designed to give priority to the most likely hypotheses. In this way, processing effort is saved in that only the most promising words are formed into possible word-sequences, and only the most promising of these are formed into phrases.

Knowledge sources may write hypotheses at various levels on the blackboard. For example, at the 'phrase' level a knowledge source called PREDICT suggests all the words that can come before and after a particular sequence of words and writes relevant hypotheses at the same 'phrase' level. A CONCAT knowledge source operating at the same level attempts to form longer word-sequences from shorter ones. It may write hypotheses at the 'phrase' and 'word-sequence' levels. Some knowledge sources use a 'bottom-up', others a 'top-down' approach. For example, hypotheses about word-sequences may be built up from evidence relating to syllables present, or different word-sequences may be evaluated on the basis of expectations generated by breaking down the meaning of a sentence into possible alternative word-sequences.

The blackboard, then, is a constantly changing pattern of hypotheses. The knowledge sources work independently of each other, responding to hypotheses and creating new ones, which in turn create new patterns on the blackboard.

HEARSAY II, described above, has now been developed into HEARSAY III. This system is not limited to speech understanding but is designed to cope with similarly complex problems in other subject areas. One particularly interesting development is that HEARSAY III has two blackboards, one to deal with constantly changing hypotheses about the problem being tackled, the other to deal with *scheduling*—managing how the system tackles the problem, in terms of deciding which hypotheses seem most promising and which knowledge structures should go ahead and when. This scheduling blackboard itself has specialized knowledge sources, which have specialized knowledge about which features of the situation are most important in making the best use of HEARSAY III's resources.

EXPERT-EASE: learning by induction

Using inductive inference a computer can, when presented with examples, discover underlying rules which even the expert who gave the original examples may not have known. All the expert has to do is to provide the examples and suggest which factors are likely to be relevant in searching for rules. The computer will do the rest. EXPERT-EASE, from Intelligent Terminals Ltd, is a commercially available system for microcomputers which does just this. It is a system which can learn from examples given to it by experts in almost any area of knowledge. It is rather like an 'automated apprentice', learning from its masters. Once it has become expert itself it can be used by non-experts to achieve expert performance levels.

What the expert does is this. First, he or she describes what factors in his or her view are important in the particular area of expertise being examined. It doesn't matter if he or she is not sure whether or not a factor is important—the system will eventually sort this out. If you think back to the 'fruit-producing' example in Chapter 11 the relevant factors—or *attributes*—are root-length; root-texture; and root-growth. The expert then gives examples in the form of different lists of *values* for the attributes. Staying with our example, values would be (for root-length) short, medium, and long; (for root-texture) smooth and hairy; (for root-growth) fast and slow. Finally, the expert would give a list of examples along with the result or *decision-class* of each example—in our example, good fruit or poor fruit. More decision classes may be included—the system is by no means limited to two. EXPERT-EASE will try to find the simplest rule which will result in its coming to correct decisions. As more examples are given, these rules are refined to take account of them.

During this process, EXPERT-EASE can provide the expert with helpful information, such as whether any of the attributes thought to be important are in fact turning out not to be so. The end-result of this learning process is a set of rules in the form of a decision tree.

Having induced rules from the examples given to it, EXPERT-EASE goes on to create automatically a consultation system, an expert system which will give advice on the basis of the rules it has learned. The designer of the system can specify the words EXPERT-EASE will use to ask questions and give advice. In our fruit-producing example if the following decision tree had been induced:

```
length
  short : texture
      smooth : poor
      hairy : growth
          fast : good
          slow : poor
  long : texture
      smooth : growth
          fast : good
          slow : poor
      hairy : good
```

the consultation system could be instructed to ask questions in a 'user-friendly' way as follows:

System: Tell me about your plant's root length
(1) Short
(2) Long
User: 2
System: How about the root texture?
(1) Smooth
(2) Hairy
User: 1
System: How about root growth?
(1) Fast
(2) Slow
User: 1
System: My conclusion is:
Your plant has an excellent chance of producing good fruit.

However, generating rules by induction is not completely dependent on having a handy expert around. For example, records can be kept of many factors associated with, say, complex factory production processes. These records could take the form of readings from meters, temperatures, valve positions, etc. Examples of different combinations of settings, readings, etc. can be obtained and classified in terms of different levels of productivity. In other words, examples, from which rules can be generated, can be obtained from recording the workings of a complex system rather than a human expert. Using EXPERT-EASE, one firm in the United States increased the volume of its business by more than $10 million a year. It is claimed that this would not have been possible without the use of EXPERT-EASE.

EXPERT-EASE is not limited to any particular area of knowledge. It has already been used in a wide variety of applications, including management decision-making, medical consultation, fault diagnosis, and others. Using the EXPERT-EASE techniques, medical experts in Yugoslavia developed a classification relating to lymphatic cancer better than had been previously achieved by human experts. A diagnosis system has been developed from this which allows non-experts to make expert diagnoses.

The system can handle up to 31 attributes, each having up to 32 'logical' values (for example, long/medium/short, hairy/smooth, etc.) or 'integer' (numeric) values from −32 766 to +32 767. Depending on the number of attributes being used, some 300 examples can be included on a personal computer—although in theory EXPERT-EASE could cope with more than 30 000!

However, files relating to different subproblems of a larger problem can be created and stored separately—up to 17 such files on a data disk. These files can then be 'chained'—linked together to produce a system capable of handling the larger problem. For example, a point in a decision tree, instead of giving a conclusion, can call up another file which will carry on where the first one left off. This file, in turn, can call up other files and so on.

More sophisticated systems offering automatic rule induction and 'translation' into consultation systems are available for larger mainframe and super-mini-machines as well as for microcomputers. For example, EX-TRAN 7 (EXpert TRANslator), from Intelligent Terminals, has been used very successfully by a number of large US and European corporations to develop expert systems. One series of four systems, designed to diagnose faults in circuit boards, has resulted in savings which run into millions of dollars. Between them, they handle about 40 problems, each of which may deal with up to 75 attributes, one with up to 100 values, leading to up to 50 decision classes!

It seems that the automatic induction of rules in this way is a very promising line of development. One of the main problems facing developers of expert systems has been the great skill, time, and trouble associated with trying to 'extract' from an expert the knowledge he or she uses in performing at an expert level. Machine-induction makes a major contribution to overcoming this problem. Often, experts solve complex problems and perform skilful actions by intuition—'by the seat of their pants'. We say that some people have a 'feel' for something, or do things 'by instinct'. Computer systems which can work out the rules underlying such instinct, feeling, intuition, would seem to have great potential.

However, imagine the potential of using automatic induction not to learn from human performance but to learn from and improve the performance of other expert systems! A variety of automatic learning techniques are used by our next example of working systems, called AM, which has produced some very impressive results.

AM and EURISKO: automated discovery using frames

Working at Stanford University in the United States, Douglas Lenat has devised computer systems capable of discovering concepts, at first in the area of number theory in mathematics and geometry, and more recently in wider areas, including the design of electronic integrated circuits, designing naval battle-fleets in a 'war-game', and modelling biological mutations in evolution. The system has proved itself capable of making genuine, non-trivial discoveries in highly complex and difficult subjects.

Lenat's Automated Mathematician (AM) uses a number of ideas we have already met in previous chapters—frames, production rules, an evaluation function to help guide the searching of the problem space, and learning by inductive inference. Basically, AM starts with a knowledge-base of 115 elementary mathematical concepts which it is given. When it runs, the system finds examples of these concepts, then, based on an examination of these examples, it generates *new* concepts and proposes new relationships between concepts. Once given the initial 115 concepts, the system runs without any human intervention, so in this sense the discoveries it makes are its own—even though it rediscovers many concepts already known to mathematicians.

AM's performance is impressive. It discovered for itself many well-known mathematical concepts such as natural numbers, addition, multiplication, perfect squares, odd and even numbers, numbers with no square roots, factoring, prime numbers, etc. Interestingly, AM discovered for itself arithmetic's fundamental theorem—Euclid's Unique Factorization Theorem (which says that there is only one way of factorizing a composite number into primes)—and Goldbach's Conjecture (that every even number larger than 2 is the sum of two primes). It discovered multiplication in four different ways, and came up with an entirely original idea about maximally divisible numbers. When a new initial knowledge-base of geometric concepts was added to AM it went on to discover concepts such as parallel, similar, congruent, rotation, etc., and came up with a novel and interesting geometric interpretation of Goldbach's Conjecture.

A development of the system, EURISKO, has been applied to a variety of other problem areas, in which it has also produced impressive results. EURISKO was responsible for a 'conceptual leap' in the design of three-dimensional integrated circuits. The design of these circuits is such a complex business that humans had not been able to come up with the idea suggested by EURISKO—that by adding parts to a junction to make it more symmetrical the junction could be made to perform more than one job at the same time. Similarly novel ideas have been suggested by EURISKO in the field of wargaming. In the Trillion Credit Squadron Competition held annually in the United States, competitors must design a naval battle-fleet which is then tested in a simulated battle against those of the other (human) competitors. EURISKO came up with some highly unconventional designs. As Michie and Johnston (1985, p. 110) note:

> On being presented at the game, Eurisko's bizarre fleets caused much laughter amongst the other players, followed by consternation when they were seen to be winning. Lenat is not surprised that for two years running the rules have been changed with the effect of outlawing Eurisko's successive concoctions.

AM's knowledge about concepts is stored in frames, as shown in Figure 42. As discussed in Chapter 9, frames are rather like standard-shaped containers with ready-made slots waiting to be filled. In Figure 42 the sort of information contained in each slot is underlined.

NAME:	A particular concept, e.g. 'prime numbers'
DEFINITIONS:	Definitions of the concept
EXAMPLES:	Examples of the concept
GENERALIZATIONS:	Related concepts which are more general than this concept
SPECIALIZATIONS:	Related concepts more specialized than this concept
CONJECTURES:	E.g. suggested relationships with other concepts
ANALOGIES:	Information on how examples of this concept are related to examples of other concepts
INTEREST:	Reasons which make the concept interesting
WORTH:	A numerical score for how interesting the concept is
VIEW:	Contains procedures for converting examples of one concept into examples of another concept. (More precise than the ANALOGIES slot)

Figure 42. AM's frame representation. (Based on Figure D4c–1 in AM by Thomas G. Dietterich. In *The Handbook of Artificial Intelligence,* vol. 3, Paul R. Cohen and Edward A. Feigenbaum (eds). Los Altos: William Kaufmann, Inc., 1982, pp. 438–51)

As well as information about concepts contained in frames, AM also has 242 production ('if. . .then') rules linked to the various slots in the frames. These rules are called heuristic rules, since they specify actions that AM should perform in order to carry out its functions. These heuristic rules can accomplish a variety of tasks:

Fill in a slot in the frame. Depending on which slot is to be filled, the procedures here will be different. They may concern finding examples of a concept, changing the numerical WORTH score (explained below), suggesting new CONJECTURES, etc.

Check a slot in the frame. Not only do these heuristic rules check that the contents of a slot are correct, they also look for interesting features about the slot. For example, they will check the examples of a concept and suggest new conjectures on the basis of any regularities they find.

Create a new concept. That is, add a new frame to the knowledge-base, and fill in at least the DEFINITIONS slot.

Two further tasks performed by the heuristic rules help AM to decide what to do next at any time. The combinatorial explosion of all possible things to do is enormous (see Chapter 5). In order to make things manageable, AM creates an agenda of tasks to be performed. Just as each concept is given a score for how interesting it is (the WORTH slot), each task on the agenda is given a WORTH score, showing how interesting the task is. AM always chooses the most interesting task to perform next. The two further tasks are therefore:

Add new tasks to the agenda. For example, rules that create a new concept can add to the agenda the task of finding examples of that concept: rules that fill in the EXAMPLES slot can also add the task of checking these examples.

Change the level of interestingness of a task. These rules can change the INTEREST slot by adding new reasons making a task interesting. This will affect the WORTH score and helps to guide the procedure of choosing what to do next.

The 'interestingness' of concepts and tasks is judged by means of 59 special heuristic production rules which give high interest ratings to, for example, extreme cases, or coincidences like the unanticipated satisfaction by examples of a concept of some other condition which is rarely satisfied.*

Generally, AM applies its heuristic rules in the following way. First, it will select the most interesting concept it can find (guided by the WORTH score). It will then try to fill in the EXAMPLES slot and, having done so, will check the examples. In doing this, AM will look for regularities in these examples. Based on the regularities it finds on checking the examples, AM then creates new concepts, suggests new CONJECTURES, and updates the INTEREST slot, modifying the interestingness of the concept it is working on.

Creating a new concept or conjecture is, in many ways, a question of finding a rule to explain examples—in other words, learning by inductive inference. AM has a large number of learning techniques, including those of changing constants into variables, and variables into constants, discussed in the card-game examples in Chapter 11. In these ways AM can generalize a concept (by extending the application of the concept to other ranges of examples), specialize a concept (by restricting its range of applicability), or generate a new concept through reasoning by analogy. It uses various other techniques involving inverting, projecting, permutating, and doing various other things to elements of existing concepts. Having created new concepts and conjectures, AM will then do its 'bookkeeping'—changing and updating other concepts in the knowledge-base where appropriate to take account of the new concepts and conjectures.

One of the main limitations of the AM system was that although it could discover new concepts it could not discover new heuristic rules. In other words, it could not learn to improve its performance by modifying its rules in the light of how well they were performing. The trouble was that it needed to do this in order to prevent its existing rules becoming increasingly less effective as its discoveries moved further away from the concepts stored in its initial knowledge-base.

EURISKO is a development of AM designed to get over this difficulty. It does so by making the heuristic rules into concepts which can themselves be

* The descriptions of heuristic rules are based on descriptions given in AM by T. G. Dietterich, (1982), in *The Handbook of Artificial Intelligence*, Vol. 3, P. R. Cohen and E. A. Feigenbaum (eds), Los Altos: William Kaufmann, pp. 438–51.

manipulated, as can all the other concepts in the system. Heuristic rules have been added which govern how heuristic rules can best be modified and improved. In this way, EURISKO can monitor the effectiveness of its own heuristic rules and modify them accordingly.

ACT: modelling the human mind

ACT is a computer system designed by John Anderson to model human memory and learning. In other words, it is a computer system built to function as its designer thinks, based on psychological research, that we think. Experiments have been carried out to see if the system behaves in the same way when confronted with certain tasks as humans do, which, in important respects, it does.

The system has a short-term, 'working' memory and a long-term one. The short-term working memory, like that in humans, has limited storage capacity and can handle ten items of information at a time. New, incoming information can in this way push out older information. Data are transferred from long-term memory into working memory for processing—i.e. for using in working things out, making inferences, calculations, etc. These features have been built in specifically to mirror the way in which it is thought that human memory works.

Information and knowledge is stored in long-term memory by means of semantic networks. If you recall from Chapter 1, these networks are composed of nodes and links, as shown in Figure 43. Here, 'John', 'Mary', and 'Fred' are nodes. The lines labelled 'likes' and 'dislikes' are links. The semantic networks used by ACT are very complex, and are made up by breaking up information into linguistic units—subject, predicate, object, etc.

Only parts of long-term memory are active at a particular time (as in human memory), and activation can spread from node to neighbouring node. The more nodes that need to be activated, the more time it takes, and they must be activated in this way before they can be transferred into working memory.

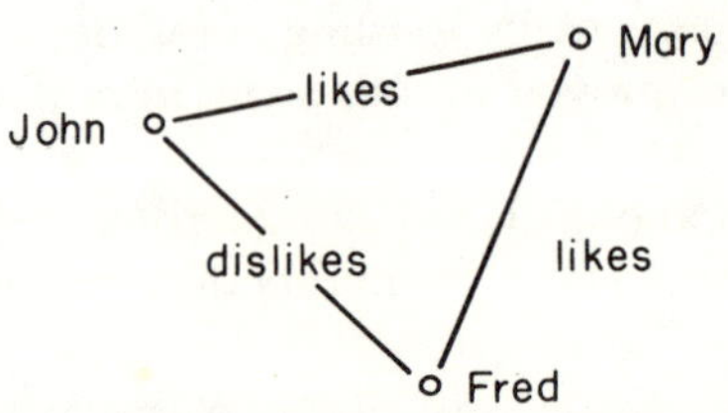

Figure 43. A semantic network

Other activated nodes, if unused in this way, are eventually made inactive again.

So much for the way knowledge is stored in long-term memory and transferred to working memory for processing. How does this processing—making calculations, making inferences—take place? ACT uses *production rules* to process the knowledge stored in associative networks. How these rules work will become clearer if we look at an example.

Imagine the task of being shown a list of numbers to remember. You are then shown another number—the *probe* number—and are asked to say whether or not it was one of the numbers included in the original list. For our example, let's assume that we are shown the list:

3 9 4 5 2

What happens in ACT (and, its creator argues, also in humans) is that an associative network is built up in long-term memory to represent the list. The structure has been simplified for the sake of clarity, and is shown in Figure 44 (a). (ACT would actually subdivide the predicates into further nodes and links.)

The production rules used in such an activity have been paraphrased below:*

Rule 1: If ACT is in the *ready* state
then keep in the ready state and put the LIST node into working memory

Rule 2: If ACT is in the *ready* state
and the probe number has been given
then change ACT's state to the *test* state, and put the probe number into working memory

Rule 3: If ACT is in the *test* state
and the list does contain
the probe number
then give message 'Found it' and put ACT into the *ready* state

Rule 4: If ACT is in the *test* state
and the list does not
contain the probe number
then give the message 'Not found' and put ACT into the *ready* state

Rule 1 will keep putting ACT in the ready state and the LIST node in working memory until any of the other rules can be applied. When the probe

* This example and the description of production rules are based on those given in ACT, by P. Cohen (1982), in *The Handbook of Artificial Intelligence*, Vol. 3, P. R. Cohen and E. A. Feigenbaum (eds.), Los Altos: William Kaufmann, pp. 50–55.

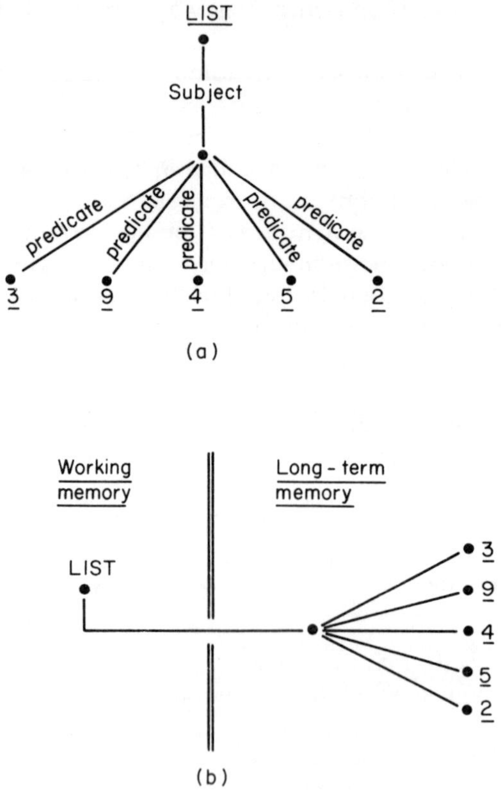

Figure 44. An associative network. (Based on Figures E3–1 and E3–2 in ACT by Paul Cohen. In *The Handbook of Artificial Intelligence*, vol. 3, Paul R. Cohen and Edward A. Feigenbaum (eds). Los Altos: William Kaufmann, Inc., 1982, pp. 50–55)

number is given, rule 2 can apply. This changes ACT's state to test, and puts the probe number into working memory to join the LIST node already there. In the test state, ACT activates the nodes joined to the LIST node—but which, unlike the LIST node itself, are in long-term memory. These nodes are 3 9 4 5 2. Once activated, they can be processed to see if the probe number is one of them. If it *is*, then rule 3 can be applied. This gives out the message that the probe number has been found and returns ACT to the ready state. If it is *not*, rule 4 is applied, giving the message that the probe number has not been found and returning ACT to the ready state.

Figure 44(b) shows ACT in the *ready* state with rule 1 being applied. The nodes connected to the LIST nodes are in long-term memory, not yet activated.

Only when ACT is in *test* state can these nodes be activated and processed to give the answer.

ACT can also learn and the simplest way to make it learn is to put in new propositions or new production rules. However, ACT can also generate new production rules. It does this by seeing if it can put, in the 'if' parts of two production rules, variables in the place of any non-variables. The learning process is similar to that described in Chapter 11. There are certain rules built into ACT to prevent it going wild and producing new production rules that are so general as to be useless! If ACT realizes, from information on the performance of its rules, that one of its production rules is too general it can produce two or more less general rules from the offending one. It does this by reversing the generalization process and replacing excessive variables in the production rule with non-variables, each of which may require a separate new rule.

New rules produced in this way are simply added to the original production rules. They do not replace them. This means that sometimes more than one production rule may be applicable—a specific rule and a more general one, for example.

ACT is equipped with a conflict-resolution rule which states that specific production rules should be applied before more general ones. Where the conflict is not between general and specific rules, ACT uses a scoring system to decide which should be given precedence. Each production rule is given a score which can change. If it is used successfully, a production rule has its score increased. If it is applied unsuccessfully (for example, it leads to a wrong conclusion) its score is decreased.

ACT is one of a number of attempts to model aspects of human thinking. Perhaps even more than the development of expert and intelligent systems designed primarily to do a job (whether or not it is done in the same way that humans do it), the prospect of increasingly successful systems designed to model human thought-processes does raise the question of whether machines can be thought of as 'intelligent', and as engaging in 'thinking'. Some would even raise the question of whether machines can be thought of as an evolving life form. Such issues are addressed in Part D.

CHAPTER 13

KNOWLEDGE-ENGINEERING TOOLS

We have just looked at some very sophisticated systems devised by very experienced artificial intelligence-workers. How can the beginner—whether a home amateur or a company wishing to explore the potential of artificial intelligence—get started? What tools are available to help not only the beginner but also the experienced researcher to build systems without constantly having to rediscover the wheel?

'Getting started' will mean different things to different people. To the individual enthusiast it may mean trying artificial-intelligence techniques on his or her home microcomputer, perhaps using BASIC. There are several good books introducing artificial intelligence for BASIC, including Brain and Brain (1984), Hartnell (1984), James (1984), Krutch (1981), Naylor (1983), and Williams (1986).

The more committed may think about buying a cheap version of PROLOG or LISP for their micro, but to the person in industry or commerce, getting started may involve buying one or more expert-system shells, possibly going on to purchase more and more sophisticated tools as their firm gains experience.

The range of languages, shells, and expert-system development tools is wide, and rapidly growing. Rather than listing them exhaustively here, I will try to show the variety of materials available and give details of where to follow up specific systems if you are interested. Costs of a number of selected systems are given in the final section of this chapter. The names and addresses of suppliers are given in Appendix 3.

Languages for artificial intelligence

It must be said that artificial-intelligence programming in BASIC is difficult. Although in theory you can do in one language anything you can do in another, the ease with which you can do it will vary. Certain languages are tailor-made for certain types of task. I would recommend to the amateur enthusiast purchasing a home micro version of PROLOG or LISP. Logic Programming Associates, for example, sell versions of PROLOG for a wide variety of microcomputers. These range from introductory micro-PROLOG 3.1, available for a number of home micros, to sigma-PROLOG for the VAX 7/80, with a range of versions between. PROLOG Professional, for example, is aimed at the software developer. It allows the use of menu options, screen windows, has good facilities for editing and 'debugging' programs, and offers a choice of PROLOG syntax.

Software can be developed then sold with a relatively cheap 'execute-only' PROLOG interpreter. MacPROLOG for the Apple Macintosh allows the use of sophisticated and very user-friendly facilities, including screen windows which may be enlarged or reduced in size, individually scrolled, and moved around the screen like a piece of paper, and excellent editing and debugging facilities. An expert-system shell—APES (Augmented PROLOG for Expert Systems)—is also available from the same firm.

Different versions of PROLOG are available from a variety of other suppliers. PROLOG 2 from Expert Systems International is fast and allows operating system commands relating, for example, to editing to be used within PROLOG programs. A version of PROLOG from Arity Corporation allows communication between PROLOG and a separate co-processor specializing in mathematical calculations (for which PROLOG was not specifically designed and which the co-processor can perform more quickly and efficiently). Salford University's LISP/PROLOG allows mixed PROLOG and LISP programming, whilst POPLOG, developed at Sussex University and available from Systems Designers (educational sales from Sussex University), allows mixed PROLOG, LISP, and POP–11 programming, and can link to other languages such as C, Pascal, or Fortran. It has good editing, online help, and teaching and demonstration facilities. Quintus PROLOG for VAX computers, from Quintus Computer Systems, Artificial Intelligence and Digital Equipment Corporation, can communicate with other languages including Pascal, Cobol, C, and Fortran. A version for Xerox LISP workstations, from Xerox Information Systems Division and Artificial Intelligence, allows a number of PROLOG programs to be run at the same time.

Versions of programming languages can range from a relatively 'raw' level—the language itself in fairly basic form—to highly sophisticated programming 'environments' which give a wide range of types and levels of support to those using them. Supporting software can be designed specifically to help the development of expert systems and artificial intelligence projects. Examples will be described below. Programming environments are particularly well developed in the case of LISP—an artificial-intelligence language that has been around for longer than PROLOG and the language used in much early and present work in artificial intelligence. As well as very sophisticated LISP programming environments, it is now possible for organizations to buy, for example, an expert system designed to teach LISP—the LISP Intelligent Tutoring System, developed at Carnegie–Mellon University, available from Advanced Computer Tutoring. The system is itself intelligent in that it carefully monitors students' progress and builds up a model or representation of each learner, which it then compares with an 'ideal' model in order to present individually tailored learning programmes.

Expert-system shells

As we saw in the case of SAVOIR from ISI (Chapter 12), it is possible to capitalize on work already done in the development of expert systems by purchasing an expert-system shell. A shell will give you in pre-programmed form an inference engine—the procedures which will draw conclusions from evidence by means of, for example, forward- or backward-chaining (or both); usually an explanation facility whereby the system will explain its line of reasoning to the user; and various other facilities, depending on the level of sophistication. What it will not give you is the expert knowledge itself, although, as we will see below, tools exist which can be extremely helpful in acquiring and making explicit such knowledge and building it into an expert system.

Expert-system shells vary greatly in what they offer. I have selected several to show some of this variety.

Quite a lot can be done using a relatively unsophisticated system employing backward-chaining and with no capability for dealing with probabilities or uncertain information. For many tasks, such facilities are unnecessary. ESP/Advisor, from Expert Systems International and written in PROLOG, is designed primarily for the purposes of 'text-animation'. This shell capitalizes on the fact that much expert knowledge is already written down in the form of text—the manuals of instructions, rules, regulations, administrative procedures, etc. that abound in business, education, government, and the professions. ESP/Advisor allows such expert knowledge to be coded in terms of specific instructions or advice on the one hand, and on the other, the conditions under which such instructions or advice should be given. The result is an interactive system able to give advice appropriate to the particular case being asked about. This seems ideal for applications in which there are complex and convoluted procedures and rules. ESP/Advisor comes complete with working examples including procedures relating to property conveyancing, taking on employees, selecting flour for baking, statutory sick pay, and working out an opening bid in contract bridge! The system also has an interface with PROLOG 1—a version of PROLOG marketed by the same firm—which allows other facilities to be developed.

Impressive results in very complex fields have been achieved by systems again employing predominantly backward-chaining but also including a capacity to handle probabilities and uncertain knowledge. MYCIN is such a system—based on 'if–then' production rules—which has produced impressive performance in diagnosing and recommending treatment for bacterial blood infections. Developed at Stanford University in the early 1970s, MYCIN is one of the best-known expert systems. In tests it has outperformed human medical experts. MYCIN is particularly interesting since from the system was developed EMYCIN (Empty MYCIN)—an expert-system shell which can be used for other application areas. EMYCIN contains the same inference engine, explanation facilities, and ability to deal with uncertainty. Other expert systems have been developed using EMYCIN in fields including income and transfer tax planning (TAXADVISOR), diagnosis and treatment of disorders of the lungs (PUFF) and blood disorders (CLOT), and advice-giving relating to pregnancy (GRAVIDA).

M.1, from Teknowledge, is a PROLOG-based (now available in C) EMYCIN-like shell for microcomputers, designed mainly to allow developers to familiarize themselves with knowledge-engineering techniques and build small prototype systems which may allow the feasibility of using these techniques to be assessed. LISP-based S.1 (now available only in C) from the same company, is a larger version of M.1, including a number of additional facilities.

Personal Consultant, from Texas Instruments, also started out as a LISP-based EMYCIN-like shell for microcomputers. It is now available in an enhanced form—Personal Consultant Plus—which extends the original 'production-rule only' way of representing knowledge to one in which frames may also be used. The system can link to other databases, and developers can easily move from the shell to the LISP language to extend the system's capabilities. Personal Consultant Plus also allows the use of 'metarules'—rules which can be used by the system to reason about its own reasoning—for example, to select which rules should be applied in a particular situation to achieve the best results.

As MYCIN led to the development of a shell not restricted to MYCIN's original area of application a microcomputer-based shell—AL/X (Advice Language/X), written in Pascal and available from Intelligent Terminals—has been developed from an expert system designed to diagnose faults on oil rigs in the North Sea. This expert system in turn was modelled to a large extent on a previous large-scale expert system—PROSPECTOR—designed to help geologists evaluate the likelihood of finding minerals in particular locations and to find the best drilling sites.

In AL/X the different strengths of links between various rules in the knowledge-base are reflected by numerical weights specified by the person designing a particular expert system. AL/X uses backward-chaining to search for possible solutions, setting a goal, and trying to establish supporting evidence for that goal. Where there is more than one possible path of exploration to choose from, AL/X uses these weights to choose the path likely to have most effect on the goal under consideration. In other words, it first seeks potentially the most influential evidence. AL/X uses forward-chaining to update probabilities (using Bayesian statistics) throughout the knowledge-base, thus affecting the choices being made in the backward-chaining search. Information can be volunteered at any stage in a consultation—the user does not have to wait to be asked a specific question—and the implications of the reply permeate throughout the knowledge-base by the forward-chaining process. AL/X has very useful debugging facilities, which means that, to some extent, it is 'transparent'—the user can see what is happening behind the scenes as the system runs. This is particularly useful in developing an expert system.

Another system, ERS (Embedded Rule-based System) from PAR Technology Corporation, is very much like AL/X—again modelled on PROSPECTOR and available in Pascal for microcomputers. ERS facilitates gathering evidence for its reasoning processes not just from responses to questions posed to the user of the system but also from one or more databases. Exsys, from Exsys Inc., is a system which also facilitates the use of external databases and can be easily linked to other programs used to control equipment. The system is also notable in its efficient use of memory. It is reported that nearly 5000 rules can be built into a system using an IBM PC with 640K.

Knowledge-acquisition tools

Other tools have been developed specifically to help in the process of knowledge-acquisition—extracting knowledge from human experts before it can be used in an expert system. One approach has been to design very user-friendly interfaces to expert-system shells which allow people without specialist computer expertise to interact easily with the system. In this way it is claimed that experts can use the shell to develop expert systems relatively independently of specialist knowledge engineers. IBM's ESE (Expert Systems Environment), for example, allows easy use by means of features including an 'English-like' rule format, automatic error-checking when rules are being put in, and good online help facilities. The system allows backward- and forward-chaining and can handle uncertain information and probabilities. User-friendliness, which allows non-computer experts to use expert-system tools relatively easily, is an increasingly common feature of expert-system shells.

Another approach is to use machine-induction. We looked at EXPERT-EASE, available from Intelligent Terminals, in Chapter 12. This system automatically identifies rules inherent in examples, whether of human experts' decisions and judgements or of the performance of complex processes. It can discover rules underlying expert decision-making and judgement even if these rules are not explicitly known by the experts themselves. This is a powerful way of acquiring expert knowledge, bypassing much of the time and effort needed to tease out expertise on the part of a human knowledge engineer (sometimes referred to as the 'Feigenbaum bottleneck'). Rules can also be discovered which might not have been discoverable in other ways. EXPERT-EASE goes on to produce a consultation system based on the rules it has identified. 'Consult-only' versions of the resultant expert system, which cannot be tampered with once the original system has been perfected, can be distributed using a low-cost version of the program. SUPER-EXPERT, also available from Intelligent Terminals, is an enhanced version of EXPERT-EASE, allowing both forward- and backward-chaining. More sophisticated systems from the same company combining these two stages of expert-system development include EX-TRAN 7 (EXpert TRANslator) and RuleMaster, one interesting application of which has been the development of Willard—a system designed to warn of severe thunderstorms!

TIMM (The Intelligent Machine Model), available from General Research Corporation, also automatically generates rules which are then built into a consultation system. TIMM can, to some extent, reason by analogy—coping with new problems by seeing how they are similar to others. Using a 'best-match' principle when faced with a problem for which all the relevant information is not available, or does not neatly fit the system's existing rules, TIMM finds those rules the 'if' parts (or conditions) of which come closest to being satisfied. In this way, new situations and incomplete information can be handled.

Knowledge-engineering environments

The dividing-line between expert-system shells and specific tools on the one hand and, on the other, general-purpose knowledge engineering 'environments' is not always clear-cut. Shells involve more pre-packaging of techniques, and tend to be committed to particular problem-solving methods and forms of knowledge-representation. As we move away from such pre-structuring, a system can become much more a tool-kit of techniques and facilities which can be used more flexibly to support a wider range of problem-solving strategies.

OPS5, available in a variety of forms from a number of companies including Artelligence, Dynamic Mastersystems, Expertelligence, and Production System Technologies, is more of a general-purpose language than a pre-packaged shell. It is particularly well suited to forward-chaining systems, having as its basic control process a loop in which rules which can be applied (i.e. rules, the conditions of which are satisfied) are constantly searched for. OPS5 has a store of production rules and a 'working memory' of facts. When the facts in working memory satisfy a number of different production rules OPS5 uses other, conflict-resolution rules to decide which rule to apply. A conflict-resolution rule may specify, for example, that a more specific rule should be applied before a more general one, or that a rule making use of more recent data from working memory should take precedence over one using older data.

The system provides a number of useful facilities designed to help the expert-system designer in his or her task. These allow the designer to trace what is going on in the program, for example, by printing out currently activated parts of the working memory; retracing steps from where a program went wrong, and 'undoing' the actions step by step backwards; and giving information as to how near particular rules are to being satisfied.

The forward-chaining facility of OPS5 (and previously OPS4) was particularly useful in the design of Digital Equipment Corporation's XCON (eXpert CONfigurer of computer systems)—with more than 4000 production rules, the largest expert system in daily commercial use. Customers ordering complex computer systems specify certain requirements. The range of possible combinations of pieces of software and hardware is enormous. An ordered system may be made up of some 400 from over 8000 possible items, different components each having up to eight characteristics which must be considered. The flexibility with which components may be configured means that there may sometimes be millions of different possibilities from which one configuration must be selected.

What XCON does is to check whether or not the customer's specification is complete or only partly specified. The customer may, for example, specify certain main requirements, leaving other details to DEC. XCON assesses this and, if necessary, supplies missing details. It checks the whole configuration for viability and produces plans specifying how all the components should be fitted together, including floor layout and cabling. XCON copes with the 'combinatorial explosion' of possible solutions be dividing up the problem into different levels of abstraction, each reflecting a different stage in its task of configuring orders. Production rules are 'partitioned' into levels, and at any one time only those rules relevant to the stage being worked on are activated.

The system is highly successful, producing acceptable configurations for 98% of the time. XCON has helped to keep DEC competitative through permitting greater flexibility and more efficiency in producing configurations, and coping with four times as many orders with the same number of human technical editors. These editors used to take about half an hour to produce configurations with an error rate of approximately 30%. XCON has reduced this error rate to 2% and produces configurations, on average, in just over 2 minutes. The system has been extended to include other facilities, including links to an ordering database and performance-monitoring.

DEC have been so pleased with the performance of XCON that they are developing a whole range of other expert systems, which include XSEL (eXpert SELling tool), which helps to configure systems and prepare quotations on customers' own premises; XSITE (eXpert SITE preparation tool), which plans computer room layouts; IDT (Intelligent Diagnostic Tool) and NTC (Network Troubleshooting Consultant), which help to diagnose problems with equipment and systems; ILOG (Intelligent LOGistics assistant), INET (Intelligent NETwork), and ISA (Intelligent Scheduling Assistant), which schedule production activity, help to maintain order flow, and assist in decision-making relating to materials-distribution.

Representing knowledge in frames is facilitated by systems including FRAME-ENGINE from Expert Systems International KEE,(Knowledge Engineering Environment) from IntelliCorp, and LOOPS, from Xerox Palo Alto Research Center. Again, these systems represent tool-kits of techniques rather than shells geared towards particular approaches. They are particularly well suited to enable 'object-oriented' programming—programming using frames, the slots of which can contain either factual information or procedures (which can include 'if ... then' production rules). Messages can be passed between frames. In this way, objects as represented by the frames become the focus of knowledge-representation. Both KEE and LOOPS allow the use of what are called 'active values'—the values of variables can be observed directly as programs run. They can also be changed and the effects of any changes observed directly. This can be particularly useful when debugging and improving a system. Smalltalk 80, also developed by Xerox at their Palo Alto Research Center, is a language and environment particularly geared to object-oriented programming. A version for the IBM PC is available from Digitalk.

Users can access KEE via microcomputers linked to a larger 'host' computer in which are stored KEE's main processing capabilities. Availability of the system is greatly increased by its ability to be accessed via telephone lines. In this way, systems can be created on a larger computer then used without the need to translate the original into a form suitable for running on the microcomputers alone. Much information is contained in frames stored in the micros which can be 'triggered' by relatively short messages from the main computer. This use of frames allows particularly efficient communication between 'host' and micro. KEE allows information from other databases to be used and enables alternative solutions to problems to be compared and, if feasible, combined. Possible solutions can be worked out and stored separately, then compared. KEE maintains a list of mutually incompatible facts which it uses to decide whether such alternative solutions can be combined. It allows 'non-monotonic reasoning'—something we met in Chapter 7 without calling it this! Non-monotonic reasoning is reasoning which enables beliefs to be revised as the pattern of evidence changes. What may have been considered true at one stage may, in the light of changing evidence, be retracted. As we saw in Chapter 7, 'non-monotonic reasoning' may be facilitated by 'dependency-directed backtracking'. KEE uses dependency-directed revision of conclusions. If a piece of information is retracted, then KEE automatically retracts any conclusions which depended on that piece of information.

Another system which allows non-monotonic reasoning is DUCK, from Smart Systems Technology, which enables systems to be developed within a LISP environment. It allows assumptions to be made provisionally in response to uncertain information. As more information becomes available, the whole knowledge-base is updated, and assumptions and conclusions retracted or modified accordingly. DUCK uses both backward- and forward-chaining in combination in such a way that the number of possible solution paths being explored can be very effectively reduced in the light of evidence and conclusions acquired as the consultation progresses. The system also allows hypothetical 'what if ...' reasoning. In other words, the user can ask what would happen if certain facts or conclusions were different. Smaller expert-system shells, such as Xi from Expertech, written in PROLOG 2 from Expert Systems International, can also allow 'what if ...' questions to be posed at the end of a consultation without the user having to go through the whole process of consultation again answering differently. Xi also uses backward- and forward-chaining and has a number of very useful facilities, including tutorial materials designed to teach the basics of expert systems and the use of Xi in particular.

ART (Automated Reasoning Tool), from Inference Corporation and Ferranti Computer Systems, offers a facility less commonly found in expert-system development tools—the use of opportunistic reasoning by means of blackboard techniques, which we looked at in relation to HEARSAY in Chapter 12. These techniques allow several relatively independent expert knowledge sources to interact, suggesting hypotheses for the group to consider, and weighing the evidence supporting each one, very much as a panel of human experts might behave in group problem-solving. The system allows backward- and forward-chaining and expert knowledge can easily be represented in the form of production rules, as in most systems, and also frames, particularly suited to problems in which knowledge is well structured and in which 'chunks' of knowledge can 'inherit' features from one another. A facility called VIEWPOINTS allows many possible solutions to be weighed up simultaneously and an optimal one selected.

Another system which uses blackboard techniques—as well as object-oriented programming—is Blobs, from Cambridge Consultants. Blobs is particularly suited to 'real-time' applications. It has a clock and can perform activities at specified points in time. It can also schedule its activities according to estimations of how long certain operations are likely to take, based on monitoring the time each 'expert' working in the system typically takes to perform its tasks.

The suitability of opportunistic reasoning and blackboard techniques for real-time applications is reflected by the use of ART in relation to space-shuttle navigation and of Blobs in relation to air defence.

Costs and capabilities

The range, prices, and capabilities of knowledge-engineering tools and environments are constantly and rapidly changing. Inevitably much detail given in any book will be out of date before the book is printed. The only way to be up to date in this field is to get hold of journals such as *Expert Systems, Expert Systems User,* and *Artificial Intelligence Review,* which frequently contain details of current and new systems along with prices and addresses for further information, and to contact suppliers for their current brochures and price-lists. Prices quoted below are those I have to hand at the time of going to press.

Systems that are initially offered at high prices for relatively large computers are increasingly becoming available in a restricted form at a much cheaper price for a range of microcomputers. A version of OPS5 is now available under the name TOPSI, from Dynamic Master Systems from $75. Admittedly, this version is more for experimenting with rather than producing fully developed systems. However,a range of increasingly expensive versions are also available, including the enhanced version of TOPSI at $175; ExperOPS5, from ExperIntelligence, at $325; OPS5+, from ArtellIgence, from $3000 to $6500; and OPS83, from Production Systems Technologies, for varying sizes of machines at prices up to $15 000. Smalltalk/V is available from Digitalk at $99.

A number of small demonstration versions of systems are becoming available for otherwise more expensive products. For example, a demonstration version of EXPERT-EASE—normally £495—is available at £20 from Intelligent Terminals. The expert system shell Insight 1 from Level 5 Research is available at $95 (the full version Insight 2 at $495). Also, there is a demonstration version of Exsis available at $10!

The National Computing Centre offers a 'starter-pack' at £650, including four systems—EXPERT-EASE, Micro Expert, ES/P Advisor, and MicroSYNCS. The first three are demonstration versions, lacking the capacity of the full systems. The expert system shell APES is available from Logic Programming Associates in two versions, APES 1 at £65 and APES 2 at £350 (with discount if purchased with a version of PROLOG).

ESP/Advisor, available from Expert Systems International, at £595 is offered to educational establishments at £295. Expert 4 is available from Elsevier Biosoft at £80.

Cheap versions of languages are also available. Turbo PROLOG is available at $99.95 from Borland. Introductory micro-PROLOG 3.1, from Logic Programming Associates, is available from £49.50 to £65 for a variety of micros.

PROLOG Professional is available from £245 to £495, MacPROLOG at £295. Logic Programming Associates' range goes up to sigma-PROLOG for the Vax 7/80 at £2000. PROLOG-86, from Solution Systems, is available, including online help and tutorial features, at $125. PROLOG 2 is available from Expert Systems International in various versions, from Personal at £145 to Professional Plus at £1395. A demonstration version is also available at £20. Systems can be expensive, of course. Quintus PROLOG can cost $17 000, for example, and the full versions of ART $85 000. The price of a system may also vary greatly, depending on the version. EX-TRAN 7, for example, from Intelligent Terminals, is available from £1995 for small machines to £29 000 for IBM mainframes. Ready-made expert systems can also vary greatly in price, from many thousands of pounds at one extreme to systems such as those available from Intelligent Environments, designed to give advice on statutory sick pay (at £49) and the Data Protection Act (at £23).

The capabilities of systems are increasing rapidly. Systems available for personal computers can handle increasingly large knowledge-bases. For example, running on an IBM PC, Exsys from Exsys Inc., and Crystal from Intelligent Environments, can handle some 5000 rules—Insight 2+ from Level 5 Research, some 8000 rules. Xi from Expertech, and KDS—a system including rule-induction, from KDS Corporation—allow very large numbers of rules to be used by linking together separate knowledge-bases. Expert systems developed on relatively large expensive tools can also be compiled—'compressed' into very efficient programs—and made available in this compiled form to run in relatively cheap forms on smaller microcomputers. The range of tools is also rapidly increasing. Some idea of this can be obtained from sources such as the *Catalogue of Artificial Intelligence Tools* (Bundy, 1984).

Another interesting development is the availability of expert systems on portable, battery-powered microcomputers. ES/P Advisor is available for the Data General One machine, for example. Goodall (1985) describes an expert system designed for use on a portable computer by firemen and other officials at the scene of a road-tanker crash. The system helps in deciding what chemicals are involved and how best to cope with them. Portable expert systems would also be of immense benefit in other contexts—for example, in getting high levels of medical expertise to those needing it, whether in industrially developed or Third World countries, particularly in times of emergency, via less expert, perhaps volunteer helpers.

Future developments

The relative 'person-hours' required to produce working expert systems have declined dramatically over recent years due to increasingly sophisticated and effective development tools. We are currently witnessing a proliferation of small systems in a wide range of application areas. However, the groundwork is also being laid in preparation for a second generation of intelligent systems. These systems will be based on large-scale, multifaceted development tools offering a wide range of integrated techniques; on improved machine-learning whereby systems can learn how to develop and improve themselves; on a greater understanding of and ability to model human information-processing and problem-solving; and on more powerful computers capable of 'parallel processing', carrying out many instructions at the same time—possibly giving in the future a capability of carrying out a billion inferences per second! Such systems will be able to handle vast quantities of data at acceptable speeds. Imagine the potential of systems able to handle such vast amounts of information and to improve their own performance by learning not only from human experts but also from monitoring their own performance and that of other expert intelligent machines.

We should not confuse the many current small-scale intelligent systems with the sort of intelligence which may be developed (or may develop) in the future. A decade ago, one of the pioneers of artificial intelligence research, Joseph Weizenbaum, stated (1976, p. 210):

> When I say therefore that I am willing to regard such a robot as an 'organism', I declare my willingness to consider it a kind of animal. And I have already agreed that I see no way to put a bound on the degree of intelligence such an organism could, at least in principle, attain.

The implications of this quotation bring us to the final part of this book, which deals with what, at first sight, seems a ludicrous question: Can, or could ever, computers be considered as intelligent life-forms?

Part D:

COMPUTERS AS INTELLIGENT LIFE?

In this final part, we move away from the practicalities of programming techniques, working systems, and development tools to a more theoretical view of computers and their future. For the development of computer systems which perform actions which, if performed by humans, would require intelligence—and the fact that we can now talk of 'artificial' intelligence—raise interesting questions. Are we just playing with words? Can we really build intelligence into machines? If so, is this intelligence fundamentally different from human intelligence? What of the future? If human and artificial intelligence are very different, can and will this difference be reduced as computer hardware and software are improved (with the help of computers themselves)? Could computers ever become *more* intelligent than humans? What relationship is there between intelligent thinking on the one hand, and life itself on the other? Could we ever consider a future computer system to be, to some extent at least, alive?

THINK!

- Answer the following questions by ticking the appropriate boxes.
-

	True	Depends	False
An aeroplane can fly			
A bird can fly			
A fish can swim			
A ship can swim			
A computer can imitate aspects of thinking			
A computer can imitate aspects of intelligent behaviour			
A computer can think			
A computer can behave intelligently			

The difference between a flying aeroplane and a swimming ship illustrates a basic difficulty in talking about artificial intelligence—the problem of words and their use. No-one would suggest that, since we use the same word to describe what a bird and an aeroplane do, they are doing exactly the same thing! Nor can we assume that computers are 'intelligent', because we use the same word to describe what they do as we use to describe human intelligent behaviour.

It is true that computer systems can imitate, or simulate, certain aspects of thinking and intelligent behaviour and perform actions and achieve ends which, if performed and achieved by humans, would require intelligence. We have seen a number of such systems in Chapter 12. For this reason you probably had little difficulty with the first six questions. You probably had more difficulty deciding on your answers to the last two—deciding whether a computer can think and behave intelligently.

In this case we cannot solve the problem merely by pointing to the ambiguity of the words involved. There is a more basic problem.

THINK!

- Now try the following questions.
- When an airline pilot switches over to the automatic pilot, is the automatic pilot

(1) *Imitating* steering the plane; *or*
(2) Actually *steering* the plane?

When a doctor types in a patient's symptoms into an expert system and receives a diagnosis, has the expert system

(1) *Imitated* diagnosing the illness; *or*
(2) Actually *diagnosed* the illness?

When a computer system makes inferences based on knowledge and rules of how to process that knowledge, is it

(1) *Imitating* thinking; *or*
(2) Actually *thinking*?

If you are anything like most of my friends you will probably readily accept that an automatic pilot is steering the plane 'for real'—but be decidedly uneasy about accepting that a computer can actually think. You may be only too ready to accept that a computer can imitate thinking—but actually think ...? That's another matter!

Where does imitation end and really doing begin? A child may learn a language by first imitating the sounds it hears, gradually associating meanings with various different words. However, precisely where does imitation end and real use of words begin? If there is no clear-cut line separating imitating from actually doing, and if one can gradually lead to the other, then can we really say with any certainty that computers cannot think?

This is an important point. It is quite something to say that a computer which can outperform human experts in complex intellectual tasks, or discover and create new knowledge previously unattainable by human beings alone—*cannot think*.

It is tempting to dismiss the question, saying that it is obvious that man-made machines cannot think. However, if we are intelligent beings we must ask: why is it obvious? If it is so obvious, then the reasons why it is so should be pretty easy to find. Remember—people thought that it was obvious that the Earth was flat and that the sun revolved around it. If we are to avoid simply playing around with words, we must grasp the nettle and ask: if computers cannot think as humans do, precisely what is that human thought has got that machine 'thought' has not?

THINK!
- Rather a lot of THINK boxes in this chapter—but this could not be more appropriate in view of what we're talking about!
- List on a sheet of paper any features of *human* thinking that no computer can have. Spend some time on this—it is an important question in our discussion.

When I tried this question on a number of my friends they came up with a list of features which they claimed were characteristic of human thinking and not of computer 'thinking'. Two features mentioned were *creativity* and *originality*. Indeed, anyone who has read any computer-generated stories or poetry might think that we have found the Achilles' heel. To date, computer systems have hardly been impressive in their creation of literary masterpieces! Many would regard the idea of specifying *rules* for producing creative and original work less than promising.

However, computers have arguably been highly creative in a number of activities, both scientific and artistic. For example, Lenat's AM system rediscovered by itself many mathematical principles and came up with some very interesting *new* ideas. Computer systems have enabled knowledge that was only implicit—chess strategies which experts were able intuitively to put into practice but never into words—to be codified explicitly. They have also synthesized new knowledge. The results of these computer programs are surely highly original and creative. 'Creative' often refers to artistic endeavour and, despite the lack of progress in the field of story- and poetry-writing, computers have been active in music and art. A key question relating to artistic creation is: to what extent is a creative artist working with and against *rules* (however intuitive or implicit, rather than explicit, or even necessarily conscious)? Surely concepts such as 'balance', 'harmony', 'structure', 'form', 'development' imply order as opposed to disorder? Concepts such as 'discord' and 'tension' imply some sort of order, insofar as harmony is not the same thing as discord. Without some rules to say what each is and is not, there would be no difference between the two. Randomness and chaos may also be involved in artistic creation. Yet even these involve some rules.

> THINK!
> - Is the following set of numbers *random*?
>   2 4 6 8 10 12 14 16 18 20 22 24 26 28 30
> - Of course it is not.
>   Now *you* write a sequence of numbers that *is* random.

How did you do it? You obviously avoided putting sequences like 5 4 3 2 1; 3 6 9 12 15 18; or 12 24 36 48 60 72. So you must have been following some rule—even if the rule was simply 'avoid any systematic order'.

I am not arguing that we know—or could ever know—all the rules governing artistic creation! What I am saying is that artistic creation may well be governed by rules—possibly involving tensions between order and chaos, harmony and discord, system and randomness. Computers are well able to handle rules for both order and randomness. The question of the extent to which they (and we) are capable of creating artistic works may relate more to their (and our) ability to perceive and use appropriate rules.

'Story grammars'—essentially, rules underlying narratives of different types—have been identified by linguists. Rules characterizing different styles of music have been identified, which can be expressed in computer programs. A number of artists and musicians are working with computers, not merely using them to store, retrieve, and print out or play back what the human artist has put in. Rather, the processing capabilities of the computer are being used in more of a collaborative effort involving a blend of human and machine information-processing.

We must be careful to avoid a 'heads we win, tails the computer loses' situation when assessing computer capability. If the computer generates random patterns we can say that no creative or aesthetic ability is involved. If more sophisticated rules are used we can say that the computer is only obeying pre-programmed rules put into it by a human artist. Once again, we come back to the same problem—to what extent are human artists ultimately dependent on rules passed to them, however intuitively and vaguely, by other artists and society? Such rules may be used as foundation-stones in the development of artists' own distinctive styles. At what point does the artist become autonomous, using his or her 'own' rules? Surely once again this is a matter of degree rather than clear-cut division.

An even more fundamental feature listed by the friends I asked as distinguishing between human and other forms of thought was *understanding*.

THINK!

- Think back to the example of an aeroplane being flown by an automatic and a human pilot.
- Whether or not you accepted that an automatic pilot could be said to be flying the plane—as opposed to imitating flying it—you would probably agree that the *understanding* possessed by the human and the automatic pilots was very different.
- Spend a few moments thinking about this difference. What exactly would the human pilot's understanding have that the automatic pilot's understanding would not?

Perhaps the most central difference between the two is the difference in *meaning* experienced by the human pilot. The human pilot has a vested interest in flying the plane safely. He or she knows what an error could mean in terms of personal safety and that of the passengers, and what safety means in the sense of fear and possible injury. Also, the pilot is aware of why he or she is flying—to earn money at one level; at another to transport passengers from one place to another for a whole variety of personal reasons. These reasons will be meaningful to the pilot. He or she will understand the idea of travelling to conduct business, to see friends, or take a holiday.

To the automatic pilot, on the other hand, the task of flying the plane would be no more or less 'meaningful' than controlling a simulated flight or a model plane.

For many years this question of meaning has been central to the debate amongst philosophers and artificial-intelligence researchers concerning the difference between intelligence as displayed by humans and that as displayed by machines. The philosopher John Searle has been a key figure in this debate. He argues that computers, as they exist at present, cannot deal with meanings—in other words, they cannot 'understand' anything in the sense that human beings do.

We can teach a parrot to say seemingly intelligent things—for example, 'Eight eights are sixty-four'. This does not mean that the parrot is intelligent. The words are meaningless to it. In a similar way, we can ask a calculator what eight eights are, and it will tell us that the answer is sixty-four. This is one better than the parrot. At least the calculator did the calculation! However, we would not describe the calculator as intelligent. It is certaintly capable of doing what if done by humans would require intelligence—but the symbols cannot be said to have any intrinsic 'meaning' to the calculator. So too with the automatic pilot.

Philosophers use the term *intentionality* to denote the deep meanings characteristic of human thought, the close links between what we think about and the actual experience of those things in the real world. When we think of 'hunger' we think of more than the string of letters making up the word or a dictionary definition—we know what hunger *feels* like, at least to some extent. We have the subjective experience of the things we think about. John Searle argues that computers as we know them at present do not possess intentionality—that is, they cannot appreciate the *meaning* of the symbols they manipulate so very effectively. He asks us to imagine a team of people in a room into which baskets of Chinese symbols are passed. The people have a set of instructions with them in the room and, by following these, they can put the various symbols into order. Having done so, they pass them out of the room. The baskets going *in* contain jumbled symbols: when they come *out*, the symbols are in order and they make sense. The people have thus turned meaningless jumbles of symbols into meaningful sentences by applying the rules they were given.

What is important to Searle's argument is that not one of the people in the room has any knowledge of Chinese. They haven't a clue what any of the symbols mean—they simply pick one up, find an instruction that tells them where to put that particular symbol, and pass the symbols out of the room. The Chinese message that they are passing out in this way is totally unknown to them.

He argues that computers as they exist at present are like the people in the room. A computer can process symbols given to it by human beings. The output of duly processed symbols, although very meaningful to the people outside, is totally meaningless to the computer. Computers cannot operate at the level of semantics (meanings), only at the level of syntax (manipulating symbols). Human beings both manipulate symbols (speech, words, and actions) *and* experience the meanings represented by the symbols.

THINK!
- Imagine that a computer prints onto a screen the words 'I am afraid'.
- Now imagine a novice skier at the top of a ski-jump saying 'I am afraid'.
- Few people would disagree that the words are more *meaningful* to the skier than to the computer!
- Now try to define exactly what you mean by *meaningful* in this sense. Imagine that you are explaining it to someone who has no idea. Jot down your definition in as much detail as you can.

Naturally, I can't guess the words you use to define *meaningful*, but I wonder whether you thought of some or all of the following points. The skier is in danger of something unpleasant actually happening to him or her. In other words, if what is feared happens, it will not be, as far as the skier is concerned, just like reading in a book that someone fell and was injured. The skier will experience the fall and the injury physically, not just mentally.

The fear will relate to the skier's own personal life and goals. He or she has personal reasons for not wanting to fall and be injured. One goal may be to avoid pain, which the skier will have experienced before in one form or another. Another may be to avoid being even temporarily disabled—this would have implications for job and personal activities. In other words, experience of fear is not just a 'cold', objective academic exercise, it has implications for and links with a person's personal, subjective goals, purposes, and feelings. He or she will experience these as distinct from those of other people.

THINK!

- In your view, could a computer ever experience *meaningfulness* in the sense of

  (1) Having its own 'internal' purposes, goals, and 'vested interests', as distinct from those of its creators or programmers; and

  (2) Links between these 'internal' purposes and the 'external' world (information from the outside world and what is happening physically to the computer)?

- What could these 'internal' purposes be?
  How could they be linked to the 'external' world?

One candidate for the role of 'meaningful' goals for a computer would be the question of internal efficiency. Programs can achieve the same ends very inefficiently, involving lots of (strictly unnecessary) processing of information. Alternatively, they can achieve the same ends more economically in terms of processing.

Computer systems exist which, when given a program produced by a human programmer, will 're-write' aspects of it in a much more efficient form. Such a system has goals which, in one way, seem very relevant to the computer, as opposed to human beings using it. The problem is that ultimately such computer-related goals are only meaningful in human terms. The whole idea of efficiency, and saving processing energy and time, is designed to benefit the people using the system. So what might seem to be the computer's 'own' internal goals are really given to it in the first place by its human masters.

If we *could* regard the computer as having its 'own' goals the idea of vested interests distinct from and, where necessary, in conflict with those of its human creators and users would in no way be far-fetched. Nor, surprisingly, might be the concept of a computer having free will—a mind of its own.

A computer system equipped not only with its own goals but also with the ability to learn how to achieve them increasingly effectively could, in some ways, be said to be out of the control of its creators and users. We have seen how existing systems can outperform their creators and synthesize knowledge previously unattainable by human beings. Is it not feasible that if such powers were applied to a computer's 'own' goals the computer might well decide to act contrary to human instructions? In other words, in certain situations advice or instructions given by its human user may be judged by the computer as harmful to its own goals and overridden. Whilst it would seem a relatively simple task to build in some ultimate human control—at the crudest level pulling the plug out!—nevertheless the computer could be said, to some extent, to have a mind of its own with its own point of view on certain matters.

The problem is the question of whether the computer could have 'own', as opposed to its users', goals. Can the goals of an artefact ever be other than given by, and meaningful only to, its human users? After all, the only reason for a computer's existence is to help human beings achieve their goals.

The answer seems clear. Artefacts cannot have their own goals. Until we turn the question round. To what extent can *we* have our 'own' goals?

THINK!

- Jot down a number of what you consider to be your own goals, purposes, or vested interests

  Make them as broad or as narrow as you like—for example, getting a job, enjoying yourself, helping people, etc.

When you really 'boil down' your list to its most basic level how certain are you that your goals originate from you personally? Many can be reduced ultimately to satisfying some biological function—hunger, safety, pleasure. Others are basically social or moral—honesty, self-respect, popularity.

I use the word 'ultimately' because we are concerned here with the origin of goals—the chicken-and-egg situation. Computers can have their own goals within the context of ultimately only serving human goals. We face a problem when we probe the origin of these ultimate goals. Certain basic biological goals are passed to us from our ancestors in our genes—in other words, given to us by others. Our social and moral goals are given to us—by the society in which we live or by divine revelation. However, if we are ultimately serving the goals of our species, and/or those of God, where does that leave the individual? Like the computer, the individual becomes primarily concerned with what are essentially subgoals, subordinated to ultimate goals which are given to him or her. The individual has a mind of his or her own, as does the computer, only when it comes to deciding how to *achieve* these ultimate goals.

> THINK!
> - How valid do you think is the idea that ultimately there is a similar relationship in terms of who sets goals between
>
>   God and the human individual;
>   The human species and the human individual;
>   Human beings and computers?
> - If the idea is not valid—where does it break down? Why can the relationship not be said to be similar?

Perhaps the most fundamental criticism of this idea is that God, the human race, and human individuals are all living beings. Computers are not. They are inanimate artefacts. Because of this, they cannot be said to be serving the goals and purposes of humans in the same way that humans may be said to be serving the goals and purposes of other living beings.

> THINK!
> - But how certain are you that no computer system could be thought of as *alive*?
>
>   No chance!
>   Future possibility.
>   Some systems can be said to be alive.
> - What, as far as you are concerned, would something have to *be*, or be able to *do* to qualify as being *alive*? In other words, list your own criteria for living beings.
> - Bear in mind that we are talking about living beings—not necessarily *intelligent* living beings. Your criteria must be able to recognize that even the simpler life-forms such as molluscs, plants, and even bacteria are all alive.

It will be interesting to compare your own criteria with those listed by Geoff Simons, in his provocatively entitled book *Are Computers Alive*? (Simons, 1983a). The book should really be titled *Computers Are Alive*—the reader is left in no doubt that the author is arguing this to be the case! After reviewing a number of scientists' approaches to defining life Simons concludes that four broad headings can include the main criteria. For something to be alive, (1) it must have a recognizable form and structure; (2) it must process energy (feed and excrete); (3) it must process information (adapt to changing conditions); and (4) it must be able to reproduce.

The first three criteria can be seen to be easily met by existing computer systems. They have a distinct form and structure; they take in energy in the form of electricity and give out energy in the form of heat. We have already seen how computers respond to changing situations in the form of information given to them—this is the essence of what computers are built to do. More difficulty, it would seem, is attached to the notion that computers can reproduce themselves.

It can be argued that, admittedly in a very restricted sense, to some extent computers can 'reproduce'. Increasingly computers are being used to produce computers with increasing autonomy. They are also helping to design more effective computer components than could be designed by human beings alone. New integrated circuits have been designed in this way. Simons also quotes the case of the Fujitsu Fanuc factory in Japan, in which robots are making robots with increasingly less human intervention. The necessary involvement of human beings, it can be argued, does not rule out the idea that computers can produce computers. The practice of one species helping another to reproduce is well known in the plant world—insects providing the necessary transport for pollen, for example.

However, reproduction in a more self-contained sense would be feasible if we could equip a computer system with (1) a description or model of itself, and (2) instructions as to how to produce another identical computer system, complete with internal model as in (1) and instructions as in (2). Such a capacity for self-reproduction would in some ways be analogous to that contained in DNA—the key to human reproduction.

In many ways the idea of consciousness—a quality thought by many to be a criterion of intelligent life (but not other levels of life)—is similar in conception to that of self-reproduction. For a machine to achieve consciousness in a relatively self-contained sense it would need a model of itself—or aspects of itself—which it could observe, think about, and change. Development in a sense similar to that in which human beings develop would be possible if changes in the model could bring about changes in the system itself. Such a system would be aware of itself, and able to change itself as a result of such awareness. We have already seen that computers can observe and 'think about' their own performance in order to improve it. In fact, very sophisticated systems have been developed which allow computers to monitor, control, rationalize, and improve their own functioning at various levels. Increasingly, 'automatic programming' systems are being developed which translate into efficiently and economically expressed programs the considerably *less* efficiently and economically expressed wishes of human programmers.

So far we have talked about what something must be able to *do* to qualify as a life-form. However, many would argue that, as well as being able to do things, life-forms must be *made of* certain substances. All known life-forms are based on carbon molecules.

In many ways, this is a circular argument. Being based on carbon molecules is a description of life-forms familiar to us—an observation rather than a statement of why life cannot exist in other forms. Clearly, a new form of life is not an already known form of life, so it cannot be included in a description of the latter. Equally we could say that no known animal has four legs and feathers, but this observation cannot itself prevent the possibility of such an animal developing.

Related to this question of life is the question: 'where' do our mental and spiritual processes 'live'? Some philosophers argue that we cannot equate thought with the results of thought. Our mental—including our spiritual—life is separate and distinct from our bodily existence. We find expression for our thoughts in the messages buzzing around in our brains and the actions and words that our bodies produce. However, this physical activity—including the functioning of the brain—is not exactly the same phenomenon as our mental activity. Our thoughts do not consist entirely of brain processes. They may result in them—or result from them!—but they cannot be equated completely with them.

It may be a mistake, therefore, to say that if we can produce in a computer processes similar to those of the human brain then the computer has thoughts. The computer is lacking the essential mental component and is capable only of the physical activities one step removed from genuine thought.

THINK! To what extent do you think that your thoughts are independent of your physical brain? Can all your thoughts be accounted for in terms of particular patterns of impulses in your brain? How far do you think your thoughts are limited by your thinking 'hardware'? Are these questions possible to answer? How can we think about them?

This 'mind/body' problem has exercised philosophers for a long time. A fundamental difficulty is that it is hard to see how our (psychological) thoughts are linked to our (physiological) 'hardware'—our brains. It would seem that our thoughts cannot be translated directly into physical processes such as neuron activity in the brain. We simply cannot explain how to play chess, for example, purely in the language of physiology—biological processes. We cannot explain our attitudes, feelings, goals, and purposes purely in such terms. It would seem that our subjective thoughts are irreducibly psycho-logical—they cannot be translated directly into more objective physiological language. They are to do with subjective meanings, and subjective meanings cannot be expressed in terms of objective physical processes.

However, if our subjective thoughts are psychological and our brain is physiological—how can one influence the other? How can psychological thoughts result in, or even result from, physical activity in a physical brain? How can there be any link? We have no objective answer to this 'mind/body' problem. However, it is interesting to note that more and more is being discovered about the relationship between the physical brain and our thoughts and actions. Methods of observing the pattern of activity in the brain have revealed interesting activity patterns related to particular thoughts in humans and other animals—from attention in monkeys to recognizing targets in fighter-pilots. It has also long been recognized that certain regions of the brain seem responsible for certain types of thinking and action.

However, it is still difficult to see in any detail how psychological thoughts can be represented and expressed in physical terms. Some would argue that the science of artificial intelligence is helping to illuminate this age-old philosophical and psychological question. Programming computers to behave in intelligent ways involves problems which seem in some ways analogous to the mind/body question. Computer-programming involves translating subjective (psychologically) meaningful problems and tasks into objectively specified electronic terms—circuits, structures, and processes. However, the meaning of programs cannot be explained purely in electronic terms. A chess-playing program, for example, cannot be explained solely in terms of circuits and electronic messages. It must be described in psychological terms—in terms of what the players are trying to achieve.

Where artificial intelligence research and development may be able to help is in suggesting ways of representing psychological ideas in terms of structures or models which do have physical electronic parallels. In other words, certain forms and structures for representing ideas can be explained to some extent in parallel—in both psychological and physical terms. The emerging link seems to be the concept of mental representations—or models—of the world and aspects of it. Certain physiological structures seem to function as such models (not in terms of any direct 'copy' but using some other type of representation). Such physical structures, acting as models or representations of aspects of the world, operate in ways which are also describable in psychological terms—in terms of subjective meanings. The development of objectively specified computer programs which behave in subjectively meaningful ways, in which qualitatively different outcomes can be linked directly to different representational and control structures and processes, may, by analogy, be able to illuminate similar links in humans. It may become less difficult to see in detail how mind and body may be directly linked.

It may be increasingly possible to explain thought and behaviour without recourse to some ephemeral mental identity existing separate from our bodies in some unknown form and location. The 'person'—what goes to make up our unified, coherent personality and 'self'—may be just such a biologically based physical system located somewhere—or in several somewheres—in the brain. This may be some sort of 'master' system in charge of lower-level systems which look after such things as vision, language comprehension, metabolic processes, etc.

Even this question of the existence of some such master system may be illuminated by artificial-intelligence research. It may well be that there is no such system in the sense of one which knows exactly what all its subsystems are up to. Computer systems such as HEARSAY, using blackboard techniques, are working examples of federations of systems in which no one system is in complete charge, and in which no one system knows all that is going on in the other subsystems. The higher-level ones may have little idea, even incorrect ideas, about what is going on at the lower levels. The whole system works—and works in terms of producing intelligent behaviour—as the result of collaboration by a number of systems, each of which knows only part of what's going on in the system as a whole. Such a picture of federal government is also supported by psychological research which indicates the fallibility of our conscious mind in knowing accurately—let alone controlling—all that is going on in our bodies. What we think of as our conscious selves has been described as no more than our 'press secretary' giving out an often incorrect version of—and sometimes being in the dark him- or herself about—what is really going on!

All of this is pure speculation. Equally, speculation could centre on precisely the opposite point of view—that the essence of human thought will never be reducible to physical processes, and that we must retain the idea of a distinct mental existence, and indeed of a central, conscious, unified 'self' carefully co-ordinating and orchestrating the whole operation, even if some of the details are left to the unconscious!

However, the latter view cannot be taken for granted, and the alternative is hardly flattering to us, in our traditional spiritual terms—a picture of a co-operating federation of purely biological systems.

We have turned the question round from whether computers could have a mental, as opposed to purely physical, existence to whether human beings may *not* necessarily have such a mental existence. Many, myself included, may find such questions disturbing. So in many ways it comes as a reassurance to know that an even more fundamental question allows us, if we wish, to maintain our own deeply held beliefs in the matter.

This fundamental question may be applied equally to proponents claiming that machines have a mental existence, as to those claiming that humans may have a purely physical existence. It is: *how could we know*? This far-reaching question can be well illustrated at a most trivial and concrete level!

THINK!
- If someone kicks you on the leg, you know what the pain feels like!
- If you kick someone on the leg, how do you know if it hurts the person? How do you know if they experience pain as you do?

The target of your boot would probably not be slow in letting you know that he or she felt pain. They might contort their face in pain, hobble, clutch their leg, or come out with some choice language!

The problem is, all these signs that they experience pain in the same way as you would if kicked are purely physical. You don't actually experience them—you can't feel directly what the person is experiencing. You assume what it must be like on the basis of your own experience. It's the same with all mental activity. You can observe the products of thinking—you can't observe thinking. So even though you may be pretty sure that, in your own case, mental activity may exist independently of your physical brain, you can only ever have evidence of the purely physical results of such activity in relation to anyone (or anything) else.

So with our present senses, we cannot know that any other person does have a mental (as distinct from physical) existence. However, neither can we know that a computer does not. For this reason, we can't actually *know* that if a computer and a person behave in identical ways, as far as we can observe, then what they are both experiencing differs in any way—even to the extent that both may not be experiencing anything but merely performing the physical signs we can observe.

It can be argued that since essentially human quality, like intentionality, cannot be measured objectively then whether or not something possesses such a quality depends on people believing that it does. If someone believes a thing to possess intentionality then that person endows it with intentionality. The corollary of this is that if I believe that I and my fellow human beings have a mental and spiritual existence separate from the body, then, as far as I am concerned, it is so. If you do not, then, as far as you are concerned, it is not so. One thing we must agree on—our mental and physical powers are insufficient to allow either of us to prove the other wrong or ourselves right!

POSTSCRIPT

It is likely that intelligent computer systems will play an increasingly important role in many aspects of our lives, whether we like it or not—indeed whether we know it or not! More and more, decisions which affect us will be taken with the help of, in consultation with, and even by computers. We will be communicating with computers: computers will be communicating with other computers. Areas of application seem almost unlimited—education, finance, industry, government, law, defence, even artistic creation.

Computers are also having considerable impact on the study of the human mind. The more we try to get computers to accomplish tasks and perform skills, the more we learn about the nature of these tasks and skills. Over the years, writers have likened aspects of the human brain to many things—from library filing systems to telephone exchanges! Arguably, computer technology offers the closest and most useful analogy yet. Compared with previous tools for developing models of how we may think, it represents a quantum leap. Also the questions, problems, and solutions encountered in our quest to develop artificial thought are proving helpful not only at a psychological level in our understanding of the nature of human thought but even at a philosophical one in our understanding of the relationship between our mental and physical existence.

Views of our future with computers differ dramatically, even amongst distinguished researchers in the field of artificial intelligence. Towards one extreme, Ed Fredkin has suggested that future computer systems may become so intelligent that they will keep us rather like pets! After all, what we say and think may be about as interesting to the supercomputer as what monkeys think and say is to us! If we were to ask such a computer what it was thinking about, during the time it would take us to ask the question the computer could have processed so much information as to make the original question meaningless in our terms! At a less extreme level, another eminent researcher, Joseph Weizenbaum (1976), has suggested that future computer technology may be as dangerous to mankind as a powerful sports-car in the hands of an inexperienced youth.

Less threatening views have also been put forward—by equally eminent workers in the field. Donald Michie, for example, writing with Rory Johnston in *The Creative Computer* (1985, p. 214) considers that artificial intelligence (AI) may bring dramatic practical benefits to mankind:

> Conventional technology has indeed been making our environment more complex and more incomprehensible, and if it continues as it is doing now the only conceivable outcome is disaster. AI seeks to reverse that process and return technology to its proper place as the obedient yet perceptive servant of humanity. Together man and machine may then be able to subdue many, perhaps in time most, of the world's afflictions.

Margaret Boden (1977, p. 473) in her scholarly *Artificial Intelligence and Natural Man*, stresses the beneficial impact of artificial intelligence at a more philosophical level:

> Contrary to common opinion ... the prime metaphysical significance of artificial intelligence is that it can *counteract* the subtly dehumanizing influence of natural science, of which so many cultural critics have complained. It does this by showing, in a scientifically acceptable manner, how it is possible for psychological beings to be grounded in a material world and yet be properly distinguished from 'mere matter'. Far from showing that human beings are 'nothing but machines', it confirms our insistence that we are essentially subjective creatures living through our own mental constructions of reality (among which science itself is one).

Work in artificial intelligence has been going on for thirty years but only recently has it emerged into a more public arena. Arguably, current high levels of interest stem less from the 'leading-edge' development of super-intelligent technology than from the widespread take-up of more intermediate but highly practicable technology—as is the case with many currently used expert systems. Perhaps the most significant impact of this field will be realized as the result of the interaction of many interest groups—interested primarily not in the technology but in the human goals which it can help to satisfy. These must remain paramount. For the moment at least, people are not machines and machines are not people. Machines can help us to achieve our goals. Intelligent machines can help us to know how we may achieve them—even to know what goals are achievable. However, they cannot (yet) tell us what our goals ought to be.

APPENDIX 1

PROGRAMMING IN PROLOG: SOME 'WORKING' ILLUSTRATIONS

Concepts such as recursion and lists are tricky to grasp at first. One of the best ways to do so is to see programs working graphically, with the variables changing from stage to stage as the programs run. This appendix is designed to show these processes in relation to a selection of programs.

Each program has two cards (see the pocket in the back of the book). For each program the first card has slots cut so that you can see the words through from the second card. Starting with the two cards absolutely parallel and lined up with each other, you then move the second card upwards behind the first, so that new information appears in the slots of the first card. In this way, you can see the programs working, filling the variables with different values at each stage. Further slots have been provided to give you a commentary on what is happening.

The programs have been designed to show particular techniques with as much simplicity as possible. These techniques are:

Program A:	Recursion
Program B:	Lists (example 1)
Program C:	Lists (example 2)
Program D:	Route-finding

APPENDIX 2

'PROBABILITY' AND 'LOOKOUT' PROGRAMS

For each program there is a full listing followed by an example consultation (preceded by a question mark).

'Probability' program

Program listing

```
begin:–
   write('What hypothesis do you want to test?'),nl,
   read(Hypoth), nl,
   probability(Hypoth,Probability),
   write(['The probability of ',Hypoth,'is ',Probability]), nl.
begin:–
   write('I cannot deal with that hypothesis'), nl.

probability(Hypoth,Probability):–
   odds_for(Hypoth,Odds_for),
   X is 1 + Odds_for,
   Probability is Odds_for / X.

odds_for(Hypoth,Odds_for):–
   prior_odds(Hypoth,Prior_odds),
   gather(Hypoth,New_Is_In,Evidence),
   multiply(l,Evidence,Answer,Total),
   Odds_for is Prior_odds * Total.
```

```
gather(Hypoth,New_ls_ln,Evidence):-
   new_ls_ln(Hypoth,Factor,New_ls_ln),
   asserta(piece_of_evidence(New_ls_ln)),
   fail.
gather(Hypoth,New_ls_ln,Evidence):-
   collect([],Evidence).

collect(List,Evidence):-
   piece_of_evidence(X), retract(piece_of_evidence(X)),
   collect([X|List],Evidence).
collect(Evidence,Evidence).

multiply(Result,[],Answer,Total):- Total is Result.
multiply(Cumulative_score,[H|T],Result,Total):-
   Result is H * Cumulative_score,
   multiply(Result,T,Answer,Total).

new_ls_ln(Hypoth,Factor,New_ls_ln):-
   ls(Factor,Hypoth,Ls),
   ln(Factor,Hypoth,Ln),
   user_answer(Factor,User_answer),
   (User_answer > 0.5, New_ls_ln is Ls * User_answer + (1 - User_answer);
   User_answer < 0.5, A is User_answer - 1,
                      New_ls_ln is Ln * A + (1 - A);
   User_answer = 0.5, New_ls_ln is 1).

user_answer(Factor,User_answer):-
   write(['How certain are you that ',Factor,' from 10 to -10']), nl,
   read(Reply),
   User_answer is Reply / 10 .

ls(dim_lights,battery_flat,6).
ln(dim_lights,battery_flat,0.8).
ls(starter_not_working,battery_flat,4).
ln(starter_not_working,battery_flat,0.2).
prior_odds(battery_flat,0.2).
```

Example consultation

```
?- begin.
What hypothesis do you want to test?
battery_flat.

[How certain are you that ,dim_lights, from 10 to -10]
10 .
[How certain are you that ,starter_not_working, from 10 to -10]
10 .
```

```
[The probability of ,battery_flat,is ,0.827586206897]
Yes
?– begin.
What hypothesis do you want to test?
battery_flat.

[How certain are you that ,dim_lights, from 10 to −10]
5 .
[How certain are you that ,starter_not_working, from 10 to −10]
5 .
[The probability of ,battery_flat,is ,0.636363636364]
Yes
?– begin.
What hypothesis do you want to test?
battery_flat.

[How certain are you that ,dim_lights, from 10 to −10]
9 .
[How certain are you that ,starter_not_working, from 10 to −10]
6 .
[The probability of ,battery_flat,is ,0.754901960784]
Yes
```

'Lookout' program

Program listing

```
expert(job(confirm_attack),
      action_conditions([attack_confirmed(definite),
                  [radar_1(positive), radar_2(positive)],
                  attack_confirmed(possible),
                  [radar_1(positive),not(radar_2(positive))],
                  attack_confirmed(possible),
                  [radar_2(positive),not(radar_1(positive))]]),
      procedures(write('ATTACK STATUS:   '))).

expert(job(expect_attack),
      action_conditions([attack_expected(high),[international_relations(bad),
                  danger_signs(observed)],
                  attack_expected(medium), [international_relations(bad),
                  not(danger_signs(observed))],
                  attack_expected(medium),[danger_signs(observed),
                  not(international_relations(bad))]]),
      procedures(write('ATTACK STATUS:   '))).
```

```
expert(job(put_on_alert),
       action_conditions([missiles_on_alert(red),[attack_expected(high)],
                     missiles_on_alert(amber),[attack_expected(medium)]]),
       procedures(write('ALERT'))).

expert(job(advise_action),
       action_conditions([fire_missiles, [attack_confirmed(definite),
                     attack_expected(high)],
                     get_on_hot_line,[attack_confirmed(definite),
                     not(attack_expected(high)),
                     not(attack_expected(medium))],
                     get_on_hot_line,[attack_confirmed(possible),
                     attack_expected(high)]]),
       procedures(write('P R I O R I T Y   M E S S A G E '))).

lookout:- nl, ask_for_data, nl, fire_rules, nl, lookout.
lookout:- nl, write('Program finished'), cleanslate, nl, nl.

ask_for_data:- write('Type an observation, "end" to end, or "." if none'),
   nl, nl, read(Answer), not(Answer = end), (Answer = [];
     assertz(fact(Answer))),
   retract_observation.

fire_rules:-
  expert(job(J),
        action_conditions([A,B|C]),
        procedures(P)),
  checkout([A,B|C],P),
  fail.
fire_rules.

checkout([A,B|C],P):- check(A,B,P), checkout(C,P).
checkout([],P).

check(A,B,P):- not(fact(A)), check_evidence(B), assertz(fact(A)), call(P),
   nl, write(A), write(' because: '), explain(B), nl, nl,
   assertz(dependency_record(A,B)), fire_rules.
check(_,_,_).

check_evidence([H|T]):- fact(H), check_evidence(T).
check_evidence([]).

explain([H|T]):- tab(40), write(H), nl, explain(T).
explain([]).
```

```
retract_observation:– write('Type an observation you wish to retract, or "." if
   none'), nl, read(Observation), (Observation = []; (retract(fact(Observation)),
   prune_dependencies(Observation))).
retract_observation.

prune_dependencies(Observation):– dependency_record(A,B),
              belongs_to(Observation,B),
           process(Observation,A,B), fail.
prune_dependencies(_).

cleanslate:– retract(fact(X)), fail.
cleanslate.

process(Observation,A,B):– nl, retract(fact(A)),
                    retract(dependency_record(A,B)),
                 write(A), write(' : now retracted'), nl, prune_dependencies(A).
process(_,_,_).

belongs_to(Item,[Item|Tail]).
belongs_to(Item,[Head|Tail]):– belongs_to(Item, Tail).

fact(not(X)):– not(fact(X)).
```

Example consultation

```
?– lookout.

Type an observation, "end" to end, or "." if none

international_relations(bad).
Type an observation you wish to retract, or "." if none

ATTACK STATUS:
attack_expected(medium)   because:   international_relations(bad)
                                     not(danger_signs(observed))

ALERT
missiles_on_alert(amber)   because:   attack_expected(medium)

Type an observation, "end" to end, or "." if none

danger_signs(observed).
Type an observation you wish to retract, or "." if none
```

```
ATTACK STATUS
attack_expected (high)        because:  international_relations(bad)
                                        danger_signs(observed)

ALERT
missiles_on_alert(red)        because:  attack_expected(high)

Type an observation, "end" to end, or "." if none

radar_1(positive).
Type an observation you wish to retract, or "." if none


ATTACK STATUS:
attack_confirmed(possible)       because:  radar_1(positive)
                                           not(radar_2(positive))

P R I O R I T Y   M E S S A G E
get_on_hot_line                  because:  attack_confirmed(possible)
                                           attack_expected(high)

Type an observation, "end" to end, or "." if none


Type an observation you wish to retract, or "." if none
radar_1(positive).

attack_confirmed(possible) : now retracted

get_on_hot_line : now retracted

Type an observation, "end" to end, or "." if none

radar_2(positive).
Type an observation you wish to retract, or "." if none


ATTACK STATUS:
attack_confirmed(possible)       because:  radar_2(positive)
                                           not(radar_1(positive))

P R I O R I T Y   M E S S A G E
get_on_hot_line                  because:  attack_confirmed(possible)
                                           attack_expected(high)

Type an observation, "end" to end, or "." if none
```

```
radar_1(positive).
Type an observation you wish to retract, or "." if none


ATTACK STATUS:
attack_confirmed(definite)    because:  radar_1(positive)
                                        radar_2(positive)

P R I O R I T Y   M E S S A G E
fire_missiles                 because   attack_confirmed(definite)
                                        attack_expected(high)

Type an observation, "end" to end, or "." if none

end.

program finished

Yes
```

APPENDIX 3

ADDRESSES OF SUPPLIERS OF SOFTWARE DEVELOPMENT TOOLS

This list is by no means comprehensive. It is restricted to suppliers of tools mentioned in Chapter 13.

Advanced Computer Tutoring Inc.
701 Amberson Avenue
Pittsburgh
PA 15232
USA

Arity Corporation
358 Baker Avenue
Concord
MA 61742
USA

Artelllgence Inc.
14902 Preston Road
Suite 212–252
Dallas
TX 75240
USA

Artificial Intelligence Ltd
62–68 Merton Road
Watford
Herts
WD1 7BY

Borland International
4585 Scotts Valley Drive
Scotts Valley
CA 95066
USA

Cambridge Consultants Ltd
Science Park
Milton Road
Cambridge
CB4 4DW

Digital Equipment Corporation
PO Box 110
Imperial Way
Reading
RG2 0TR

Digitalk Inc.
5200 West Century Boulevard
Los Angeles
CA 90045
USA

Dynamic Master Systems Inc.
PO Box 566456
Atlanta
GA 30356
USA

Elsevier Biosoft
68 Hills Road
Cambridge
CB2 1LA

ExperIntelligence Inc.
559 San Ysidro Road
Santa Barbara
CA 93108
USA

Expertech Ltd
Expertech House
172 Bath Road
Slough
SL1 3XE

Expert Systems International Ltd
9 West Way
OXFORD
OX2 0JB

Exsys Inc.
PO Box 75158
Albuquerque
NM 87194
USA

Ferranti Computer Systems Ltd
Ty Coch Way
Cwmbran
Gwent
NP44 7XX
Wales

General Research Corporation
7655 Old Springhouse Road
MacLean
VA 22102
USA

IBM Corporation
Department 68Y
PO Box 152750
Irving
TX 75015
USA

Inference Corporation
5300 West Century Boulevard
Los Angeles
CA 90045
USA

IntelliCorp
707 Laurel Street
Menlo Park
CA 94025
USA

Intelligent Environments Ltd
20 Crown Passage
St James
London
SW1Y 6PP

Intelligent Terminals Ltd
George House
36 North Hanover Street
Glasgow
G1 2AD

ISI Ltd
11 Oakdene Road
Redhill
Surrey
RH1 6BT

KDS Corporation
934 Hunter Road
Wilmette
IL 60091
USA

Level 5 Research Inc.
503 Fifth Avenue
Indialantic
FL 32903
USA

Logic Programming Associates Ltd
Studio 4
The Royal Victoria Patriotic Building
Trinity Road
London SW18 3SX

National Computing Centre
Oxford Road
Manchester
M1 7ED

PAR Technology Corporation
PAR Technology Park
220 Seneca Turnpike
New Hartford
NY 13413
USA

Production System Technologies Inc.
642 Gettysburg Street
Pittsburgh
PA 15206
USA

Quintus Computer Systems Inc.
2345 Yale Street
Palo Alto
CA 94306
USA

Salford University
Industrial Centre
Salford
M5 4WT

Smart Systems Technology Inc.
6870 Elm Street
McLean
VA 22101
USA

Solution Systems Inc.
335–B Washington Street
Norwell
MA 02061
USA

Sussex University
Professor Aaron Sloman
Cognitive Studies Programme
Arts Building E
University of Sussex
Brighton
BN1 9QN

Systems Designers plc
Pembroke House
Pembroke Broadway
Camberley
Surrey
GU15 3XH

Systems Designers Software Inc.
Suite 407
444 Washington Street
Woburn
MA 01801
USA

Teknowledge Inc.
525 University Avenue
Palo Alto
CA 94301
USA

Texas Instruments
Manton Lane
Bedford
MK41 7PA

Texas Instruments Inc.
PO Box 809063
Dallas
TX 75380
USA

Xerox Palo Alto Research Center
3333 Coyote Hill Road
Palo Alto
CA 94304
USA

Xerox Corporation
250 North Halstead Street
PO Box 7018
Pasadena
CA 91109
USA

BIBLIOGRAPHY

Alty, J. L., and Coombs, M. J. (1984) *Expert Systems: Concepts and Examples*, Manchester: National Computing Centre. A detailed and concise introduction to concepts and techniques. Not for the absolute beginner.

Amsterdam, J. (1983), Computer languages of the future. *Popular Computing*, September, 136–51. Gives thumb-nail sketches of programming details distinguishing PROLOG, LISP, FORTH, APL, Module-2, and Smalltalk.

Arnold, W. R., and Bowie, J. S. (1986). *Artificial Intelligence: a Personal, Commonsense Journey*, Englewood Cliffs, NJ: Prentice-Hall. A wideranging overview of topics including history, languages, software, hardware, applications, and social aspects.

Artificial Intelligence. A journal from Elsevier Science Publications, Amsterdam. A scholarly journal, not for the beginner.

Artificial Intelligence Review. A journal from Blackwell Scientific Publications, Oxford. Includes reviews of products and tutorial articles, not requiring a lot of foreknowledge of the area.

Barr, A., and Feigenbaum, E. A. (eds) (1981). *The Handbook of Artificial Intelligence*, Vols I and II, Los Altos, CA: Kaufman. The three volumes (*see also* Cohen, P. R. and Feigenbaum, E. A. (eds) (1982)) provides a comprehensive, detailed description of AI concepts and techniques. An excellent reference work.

Benson, I. (ed.) (1986). *Intelligent Machinery: Theory and Practice*, Cambridge: Cambridge University Press. Not for the beginner. Describes trends in software research and directions for medium- and long-term development.

Boden, M. (1977). *Artificial Intelligence and Natural Man*, Brighton: Harvester. A scholarly discussion of concepts, techniques, systems, and broader philosophical and psychological issues of AI. Hard reading for the beginner but a classic text.

Bonnet, A. (1985). *Artificial Intelligence: Promise and Performance*, Englewood Cliffs, NJ: Prentice-Hall International. Quite detailed. Not for the absolute beginner. Covers many concepts with examples. Deals with applications in research and industry, including expert systems and natural-language processing.

Brain, K. R., and Brain, S. (1984). *Artificial Intelligence on the Spectrum Computer: Make Your Micro Think*, London: Sunshine. Other versions include the BBC and Electron, Dragon, Commodore 64, Sinclair QL, with appropriate changes in the titles.

Bratko, I. (1986). *PROLOG Programming for Artificial Intelligence*, Reading, Mass.: Addison-Wesley. An excellent tutorial introduction to PROLOG and AI applications. A 'must' for students of PROLOG.

Bundy, A. (ed.) (1980). *Artificial Intelligence: an Introductory Course,* Edinburgh: Edinburgh University Press. Lecture notes for an introductory course in AI given in 1976–7. Aimed at university undergraduates without prior experience. Requires concentrated rather than bedtime reading! Shows in detail how a number of problems can be solved by programs in LOGO and LISP.

Bundy, A. (ed.) (1984). *Catalogue of Artificial Intelligence Tools,* Berlin: Springer-Verlag. Not for the beginner, except to show the very wide range of software available.

Charniak, E., and McDermott, D. (1985). *Introduction to Artificial Intelligence*, Reading, Mass.; Addison-Wesley. For the serious student. An excellent introduction.

Clark, K. L., and McCabe, F. G. (1984). *Micro-PROLOG: Programming in Logic*, Englewood Cliffs, NJ. Prentice-Hall. A more advanced introduction than Ennals, R. (1984). *Beginning Micro-PROLOG.*

Cohen, P. R., and Feigenbaum, E. A. (eds) (1982). *The Handbook of Artificial Intelligence*, Vol. III, Los Altos, CA: Kaufman. For notes, see Barr. A., and Feigenbaum, E. A. (eds) (1981).

Conlon, T. (1985). *Start Problem-Solving with PROLOG*, Wokingham: Addison-Wesley. A good introduction to Micro-PROLOG for the beginner.

De Saram, H. (1985). *Programming in Micro-PROLOG*, Chichester: Ellis Horwood. Aimed at the beginner. Introduces Micro-PROLOG and includes sections on games, turtle graphics databases, and natural language. As well as other programs, a small expert-system shell is listed in Micro-PROLOG.

Doores, J., Reiblein, A., and Goodall, A. (1986). *PROLOG: Programming for Tomorrow*, Chichester: Sigma Technical. Aimed particularly at the personal computer-user who is familiar with BASIC or Pascal. Forthcoming at the time of going to press.

Ennals, R. (1985). *Artificial Intelligence: Applications to Logical Reasoning and Historical Research,* Chichester: Ellis Horwood. A very interesting book, deserving a wide audience. Fuses ideas about teaching and understanding history and the use of artificial intelligence. Ideas are developed in micro-PROLOG. Excellent for those interested in AI applications, PROLOG, history, and education. Although not specifically designed to teach micro-PROLOG, the book requires no previous knowledge of PROLOG, and introduces useful programming ideas, described in micro-PROLOG.

Expert Systems. A journal from Learned Information, Oxford. An excellent journal with news, product details, articles describing expert systems, and review articles. For the newcomer and the experienced.

Expert Systems User. A journal from Expert Systems User Ltd, London. Another excellent journal, more frequent than *Expert Systems*. Includes news, articles, and reviews.

Feigenbaum, E., and McCorduck, P. (1984). *The Fifth Generation*, London: Michael Joseph. A popular introduction to AI and the race to develop fifth-generation highly intelligent computers.

Forsyth, R. (ed.) (1984). *Expert Systems: Principles and Case Studies*, London: Chapman and Hall. A series of specialist chapters on particular aspects of designing expert systems. Not for the beginner.

Forsyth, R., and Naylor, C. (1985). *The Hitch-Hiker's Guide to Artificial Intelligence*, London: Chapman and Hall/Methuen. Available in BBC BASIC and APPLESOFT

BASIC versions. Also relevant to other BASICs. Introduces some key concepts in AI with listings of programs in BASIC. For those already experienced in BASIC programming.

Goodall, A. (1985). *The Guide to Expert Systems*, Oxford: Learned Information. A useful guide, particularly for those in business and industry wanting a good overview. Key concepts are introduced along with a number of examples of systems. Includes a selective list of producers of AI and expert-systems development tools. Annoyingly, the book does not have an index.

Harmon, P., and King, D. (1985). *Expert Systems: Artificial Intelligence in Business*, New York, Wiley. A useful guide especially for those in business and industry wanting a good overview of what AI has to offer. Includes comparative tables describing the features of different systems, a glossary of terms, a list of companies, and (1985) product prices.

Hartnell, T. (1984). *Exploring Artificial Intelligence,* London: Interface Publications.

Hasemer, T. (1984). *A Beginner's Guide to LISP,* Wokingham: Addison-Wesley. An excellent introduction to LISP oriented to AI applications for the beginner with a microcomputer (also useful for learning LISP on other machines).

Haugeland, J. (1985). *Artificial Intelligence: the Very Idea*, Cambridge, Mass.: MIT Press. A good book for the philosophically inclined. The central theme is how far human thinking and computer processing can be equated.

Hayes-Roth, F., Waterman, D. A. and Lenat, D. B. (eds) (1983). *Building Expert Systems,* Reading, Mass.: Addison-Wesley. Not for the beginner. A very useful review of concepts, techniques, and development tools.

Jackson, P. (1986). *Introduction to Expert Systems,* Wokingham: Addison-Wesley. An introduction aimed at the student, professional, and interested layman.

James, M. (1984). *Artificial Intelligence in BASIC,* London: Newnes Technical.

Johnson, L., and Keravnou, E. T. (1985). *Expert Systems Technology: a Guide,* Tunbridge Wells: Abacus Press. Certainly not for the beginner. Gives details of the processes used in programming a number of well-known expert systems including MYCIN, PROSPECTOR, PIP, CASNET, INTERNIST-I, etc.

Krutch, J. (1981). *Experiments in Artificial Intelligence for Small Computers,* Chichester: Sigma Technical. Examples in Microsoft Level II BASIC.

McCorduck, P. (1979). *Machines who Think*. San Francisco: W. H. Freeman. Describes the history and development of AI.

Michalski, R., Carbonell, J. R., and Mitchell, T. M. (eds) (1983). *Machine Learning,* Palo Alto: Tioga Press. Not for the beginner. A collection of specialized articles on aspects of machine learning.

Michie, D. (ed.) (1982). *Introductory Readings in Expert Systems,* New York: Gordon and Breach. A very useful collection of articles.

Michie, D., and Johnston, R. (1985). *The Creative Computer*. Harmondsworth: Penguin. Written in a journalistic way, yet gives an overview of key concepts in current AI research.

Naylor, C. (1983). *Build Your Own Expert System,* Chichester: Sigma Technical. Includes aspects of machine-learning and Bayesian inference, in BASIC.

Open University (1978). D303. *Cognitive Psychology,* Block 4, Units 26–8. *Learning and Problem Solving,* Part 3, Milton Keynes: Open University Press. An excellent introduction to key AI ideas. I believe that this course is being revised to include, amongst other things, examples in PROLOG. Often Open University course units are made generally available, as for the D303 course. When the revised course units are available, they will be a 'must' for those seeking an up-to-date introduction.

O'Shea, T., and Self, J. (1983). *Learning and Teaching with Computers: Artificial Intelligence in Education,* Brighton: Harvester. An excellent, very readable introduction to this area.

O'Shea, T., and Eisenstadt, M. (eds) (1984). *Artificial Intelligence: Tools, Techniques and Applications*, London: Harper and Row. Not for the beginner. A collection of articles on various aspects of AI.

Rich, E. (1983). *Artificial Intelligence,* New York: McGraw-Hill. Not for the beginner. Suitable as a serious text for those with computing experience and preferably some mathematics expertise. Covers many key concepts in AI. A good introduction for the serious student.

Rogers, J. B. (1986). *A PROLOG Primer,* Reading, Mass.: Addison-Wesley. A good tutorial text for the beginner.

Scown, S. J. (1985). *The Artificial Intelligence Experience: an Introduction,* Maynard, Mass.: Digital Equipment Corporation. A very good, concise introduction and overview of AI concepts, techniques, and development tools.

Sell, P. S. (1985). *Expert Systems—a Practical Introduction,* Basingstoke: Macmillan. Needs some prior programming experience. Includes a description of procedures to make a simple expert-system shell, general enough to allow programmers to produce the shell in their preferred language.

Simons, G. L. (1983a) *Are Computers Alive? Evolution and New Life Forms*, Brighton: Harvester. A thought-provoking book. Simons makes some extreme claims, but offers detailed support for most of them. Worth reading.

Simons, G. L. (1983b) *Towards Fifth-Generation Computers*, Manchester: National Computing Centre. A good introduction.

Simons, G. L. (1984). *Introducing Artificial Intelligence,* Manchester: National Computing Centre. Gives a wideranging overview of AI, approaching it in a broad context. Written in a non-technical way. Good for the beginner.

SPL-Insight Series (1986–). Wokingham: Addison-Wesley. These are forthcoming at the time of going to press. They look interesting. Some will be aimed at managers wanting an overview of developments, others at a more technically minded audience.

Stede, M. (1986). *Introduction to Artificial Intelligence*, Chichester: Ellis Horwood. Written to introduce the non-specialist to AI techniques.

Stefik, M. J. (1980). *Planning with Constraints,* PhD dissertation, Stanford University: Computer Science Department.

Stevens, L. (1985). *Artificial Intelligence: the Search for the Perfect Machine,* Hasbrouck Heights, NJ: Hayden Book Company. Overview of many AI areas including image-understanding, robotics, natural language, voice-recognition, expert systems.

Turing Institute Knowledge Engineering Tutorial Series (1986–). Wokingham: Addison-Wesley. Forthcoming at the time of going to press, but they must be mentioned here. The Turing Institute is an internationlly renowned centre for AI research and development.

Turner, M. (1984). *Expert Systems: a Management Guide*, London: PA Computers and Telecommunications. An introductory guide for non-specialist managers designed to give an overview of potential of expert systems. Only 62 pages long, but a useful little book.

Waterman, D. (1985). *A Guide to Expert Systems,* Reading, Mass.: Addison-Wesley. Written for the non-computer expert. An excellent survey of expert systems—how to build them and how to choose them. Includes details of over 200 products.

Weiss, S. M., and Kulikowski, C. A. (1983). *A Practical Guide to Designing Expert Systems,* London: Chapman and Hall. A good introduction to the practicalities of building systems. Not for the absolute beginner.

Weizenbaum, J. (1976). *Computer Power and Human Reason: From Judgement to Calculation,* San Francisco: W. H. Freeman. Persuasive cautionary words, from one of the field's pioneers, about the dangers of AI. Highly recommended.

Williams, N. (1986). *The Intelligent Micro: Artificial Intelligence for Microcomputers,* London: McGraw-Hill. AI techniques for QL SuperBASIC. Also relevant to other BASICs.

Winston, P. H. (1984). *Artificial Intelligence,* Reading, Mass.: Addison-Wesley. An excellent text covering the key concepts and techniques of AI for the serious student. Requires detailed reading.

Winston, P. H., and Horn, B. K. P. (1984). *LISP,* 2nd edn, Wokingham, Addison-Wesley. An excellent introduction for the serious student.

Winston, P. H., and Prendergast, K. A. (eds) (1984). *The AI Business,* Cambridge, Mass.: MIT Press. Discusses commercial aspects of AI.

Yazdani, M., and Narayanan, A. (eds) (1984). *Artificial Intelligence: Human Effects,* Chichester: Ellis Horwood. Examines the impact of AI applications in a broad context.

INDEX

ABSTRIPS, 136–139
ACT, 154–157
active values, 165
addresses, 195–198
AL/X, 161–162
AM, 150–154, 173
ambiguity, 90–100, 144–147
APES, 159–167
ART, 166–168
artificial-intelligence languages, 3–4, 11, 158–160, 167–168
asynchronous problem-solving, 101–117, 144–147, 166–167, 190–194
automatic rule-induction, 118–130, 147–154

backtracking, 59, 73–76, 113–117, 143–144, 165–166, 190–194
backward-chaining, 19, 92–93, 102, 143–147, 160–162
BASIC, 3, 11, 158–159
Bayesian updating, 81–89, 141–144, 162, 188–190
belief revision, 73–76, 113–117, 143–144, 165–166, 190–194
blackboard techniques, 79, 101–117, 144–147, 166–167, 183, 190–194
BLOBS, 166–167
Boolean operators, 82–83, 141–144
bottom-up processing, 19, 92–100, 102, 105–117, 143–147, 162–163
breadth-first search, 59–62

CADUCEUS, 2
certainty, 80–89, 141–144, 162–163, 188–190
CLOT, 161
combinatorial explosion, 56
conflict resolution, 157, 164
costs, 167–168
COUNSELLOR, 144
criticality values, 72, 136–139
CRYSTAL, 168

data-driven processing, 19, 92–100, 102, 105–117, 143–147, 162–163
decision trees, 124–130, 147–150
default values, 95–97
DENDRAL, 2
dependency-directed backtracking, 73–76, 113–117, 143–144, 165–166, 190–194
depth-first search, 59
DUCK, 166

EMYCIN, 161
environments, 163–167
ERS, 162
ESE, 162
ESP/ADVISOR, 160–161, 167–168
EURISKO, 150–154

expectations, 90–100, 144–147
ExperOPS 5,167
EXPERT, 4, 167
EXPERT-EASE, 147–150, 162–163, 167
expert-system shells, 141–144, 160–162
EXSYS, 162, 168
EX-TRAN, 7, 150, 163, 168

FEASA, 144
Feigenbaum Bottleneck, 163
forward-chaining, 19, 92–100, 102, 105–117, 143–147, 162–163
FRAME-ENGINE, 165
frames, 94–100, 101–104, 150–154, 161, 165, 190–194
future developments, 168–169, 185–186
fuzzy logic, 81–83

goal-driven processing, 19, 92–93, 102, 143–147, 160–162
goal-state, 57
GRAVIDA, 161

HAM (Human Associative Memory), 8
HEARSAY, 144–147, 183
heuristics, 55–79, 152–154
Human Associative Memory (HAM), 8

ID3, 123–130
IDT, 165
ILOG, 165
inductive inference, 118–130, 147–154
INET, 165
initial state, 57
INSIGHT, 167–168
intentionality, 176–179
ISA, 165

KDS, 168
KEE, 165–166
knowledge-acquisition tools, 147–150, 162–163
knowledge-engineering environments, 163–167
knowledge-engineering tools, 158–169

languages, 3–4, 11, 158–160, 167–168

learning, 118–130, 147–154
LISP, 158–160
lists, 42–45, 187
LN factor, 85–89, 141–144, 188–190
local necessity factor, 85–89, 141–144, 188–190
logical sufficiency factor, 85–89, 141–144, 188–190
long-term memory, 92–93, 154–157
LOOPS, 165
LS factor, 85–89, 141–144, 188–190

M.1, 161
machine learning, 118–130, 147–154
MacPROLOG, 159, 168
macro-operators, 134–136
MACROPS, 134–136
meaning, 175–179
means–ends analysis, 63–70, 133–139
MEMOD, 8
Micro Expert, 167
Micro-PROLOG 3.1, 159, 167
MicroSYNCS, 167
mind/body problem, 181–186
MOLGEN, 139–141
monotonic reasoning, 165–166
multidirectional problem-solving, 77–79, 144–147, 166–167
MYCIN, 17, 161

non-monotonic reasoning, 165–166
NTC, 165

object-oriented programming, 165
operators, 66–70, 133–139
opportunistic problem-solving, 77–79, 101–117, 144–147, 166–167, 190–194
OPS5, 163–165, 167
OPS83, 167

parallel processing, 169
Personal Consultant, 161
philosophical aspects, 171–186
PLANEX, 136
plans, 63–70, 94, 133–141
POP-2, 159
POPLOG, 159
probabilities, 80–89, 141–144, 162, 188–190

production rules, 17–19, 142, 150–157, 161, 164
PROLOG, 3–4, 11, 158–160, 167–168
PROLOG-2, 159, 168
PROLOG Professional, 159, 168
PROSPECTOR, 2, 161
pruning, 74–76
PUFF, 2, 17, 161

Quintus PROLOG, 159, 168

R1, 2
recursion, 21–32, 187
rule-based systems, 17–19, 142, 150–157, 161, 164
rule-induction, 118–130, 147–154
RuleMaster, 163

S.1, 161
Salford University LISP/PROLOG, 159
SAVOIR, 19, 141–144
scheduler, 147
schemas, 94
scripts, 94–97
semantic network, 7–11, 18, 154–157
semantics, 176–179
SHAKEY, 133–136
shells, 141–144, 160–162
short-term memory, 92–93, 154–157
SIGMA-PROLOG, 159, 168
slots, 96–99, 102–104, 150–154, 165, 190–194
SMALLTALK, 165, 167
Starter Pack, 167
state space, 59
STRIPS, 133–139
super-EXPERT, 163
super-operators, 134–136
suppliers, 195–198
syntax, 176–179

TAXADVISOR, 161
Teachable Language Comprehender (TLC), 8
text animation, 160–161
TIMM, 163
TLC (Teachable Language Comprehender), 8
top-down processing, 19, 92–93, 102, 143–147, 160–162
TOPSI, 167
triangle tables, 134–136
TURBO-PROLOG, 167

uncertainty, 80–89, 141–144, 162–163, 188–190

VIEWPOINTS, 166

WILLARD, 163
working memory, 92–93, 154–157

XCON, 164
Xi, 168
XSEL, 165
XSITE, 165